Animal Clinical Chemistry

A Primer for Toxicologists

Animal Clinical Chemistry

A Primer for Toxicologists

Edited by
G. O. EVANS

Taylor & Francis
Publishers since 1798

UK Taylor & Francis Ltd, 1 Gunpowder Square, London EC4A 3DE
USA Taylor & Francis Inc., 1900 Frost Road, Suite 101, Bristol, PA 19007

British Library Cataloguing in Publication Data

A catalogue record for this book is available from the British Library

ISBN 0-7484-0350-7 (cased)
ISBN 0-7484-0351-5 (paperback)

Library of Congress Cataloguing in Publication data are available

Cover design by Amanda Barragry Design

Contents

Preface

In 1975, clinical chemists from several pharmaceutical and contract toxicology laboratories met to discuss problems of mutual interest and thus the Industrial Clinical Chemistry Discussion Group was established in the United Kingdom. The group changed its name to the Animal Clinical Chemistry Association in 1980, and reaffirmed its primary aim of advancing the science of animal clinical chemistry in safety evaluation, toxicology and veterinary science. The Association promotes national and international meetings and as part of its educational activities, several review papers have been published and a training course established. The authors in this book have been involved in many of these various activities and I am pleased that again they have made their enthusiastic contributions. The Editor thanks the staff members at Taylor & Francis and their agents for their assistance in the completion of this book.

G. O. Evans

Contributors

D. T. Davies
Safety of Medicines Group, Zeneca Pharmaceuticals, Macclesfield, SK10 4TG, UK.

A. Dickens
Information Services, Unilever UK Central Resources Ltd, Bedford, MK44 1LQ, UK.

G. O. Evans
Safety Assessment, Astra Charnwood, Loughborough, LE11 0RH, UK (previously with Wellcome Research Laboratories, Beckenham).

J. Robinson
Environmental Safety Laboratory, Unilever UK Central Resources Ltd, Bedford, MK44 1LQ, UK.

M. D. Stonard
Central Toxicology Laboratory, Zeneca Ltd, Macclesfield, SK10 4TJ, UK.

D. D. Woodman
Toxicology, SmithKline Beecham Pharmaceuticals, Welwyn, AL6 9AR, UK.

M. J. York
Pathology, Medical Safety Evaluation, Glaxo-Wellcome plc, Ware, SG12 0DP, UK.

1

General Introduction

G. O. EVANS

> Toxicology is the discipline that integrates all scientific information to help preserve and protect health and the environment from the hazards presented by chemical and physical agents.
>
> (Society of Toxicology: Miya *et al.*, 1988)

During our lives, we encounter a wide range of xenobiotics (foreign chemicals): these chemicals include drugs, manufactured consumer products including food additives, environmental pollutants, pesticides, industrial chemicals, and naturally occurring substances. Xenobiotics may be organic or inorganic chemicals of both synthetic and natural origin, and they also include those substances now loosely referred to as 'biologicals', e.g. vaccines and monoclonal antibodies. In recent years, there has been a significant growth in the international activity of safety testing to meet the increasing requirement for consumer safety and to deal with the number and wide range of compounds.

In vivo studies enable toxicologists to make predictions for the safety of xenobiotics in man and other species, and in this book we will be concentrating on laboratory tests used in safety evaluation studies with animals. To predict risk factors associated with the test compound(s), data from clinical chemistry tests require interpretation and integration with the other data obtained from toxicological studies. In our discussions, we will not be including clinical toxicology which may be defined as the analysis of drugs, heavy metals and chemical agents in body fluids and tissues (relating to the management of human or animal medicine), nor shall we be dealing with aspects of forensic toxicology and the related subject of analytical toxicology.

For *in vivo* safety evaluation studies Zbinden (1993) proposed three principal goals which are:

1 Spectrum of toxicity – detection of adverse effects in selected laboratory animal species and description of the dose–effect relationship over a broad range of doses.

2 Extrapolation – prediction of adverse effects in other species, particularly man.
3 Safety – prediction of safe levels of exposure in other species, particularly man.

For food additives, often with low biological activity, the onus on the toxicologist is to demonstrate a 'no observable adverse effect level' (NOAEL), i.e. a non-toxic level, and to identify separately any changes in data which may be the result of an adaptive response to repeated overdosage. This is in contrast to the situation with drugs, where it is important to both demonstrate potential toxic response and to define a 'no observable effect level' (NOEL): here the difficulties are in distinguishing between non-adverse pharmacological response(s), desired pharmacological action(s), adaptive response(s) and apparent toxic effect(s) (James, 1993). It remains debatable whether safety evaluation should be based on demonstrating the absence of toxic signs rather than target organ toxicity (Heywood, 1981). Toxicology studies are generally designed to characterize the adverse effects of a xenobiotic by identifying its effects on target organs and on key metabolic functions, and furthermore to determine if these effects are reversible.

Many of the procedures used in conventional toxicity testing have arisen by an apparently empirical process, e.g. period of exposure and selection of dose levels. Safety studies are performed with a variety of experimental animals including rodents, rabbits, non-human primates and farm animals before progressing to studies with human volunteers, and other animals in the case of veterinary medicines. The individual study designs vary with the chosen animal species, route of administration, duration (dependent on the proposed or estimated exposure in man), and dosages which may represent multiples of the proposed therapeutic dosage or different estimates of environmental exposure.

Test compounds can cause both local and systemic adverse effects. Local effects occur at the sites of initial contact, e.g. at injection sites, while systemic effects are observed in organs remote from the site of initial exposure. The degree of toxicity is dependent on the number (or duration) of exposures. Effects can be classified as acute, i.e. short-term following a single dose, while repeated or multiple exposure effects can be referred to as short-term (not more than 5% of life span), sub-chronic (5 to 20% of life span) and chronic (entire or major part of life span).

Often the distribution and metabolism of compounds vary between different animal species, and equal concentrations of chemical (or metabolites) in different species do not mean equal pharmacological or toxic effects. Strain and sex differences in the metabolism of xenobiotics are also well recognized (Timbrell, 1991). The selective applications for insecticides and pesticides make use of these variations between species. Toxic effects may be caused not by the parent compound but by one or more metabolites, and sometimes the identity of these metabolites in different species is not known at the time of animal toxicology studies. The development of pro-drugs, designed to overcome pharmaceutical or pharmacokinetic limitations associated with the parent drug molecule, presents a different challenge, for although the pro-drug may be pharmacologically inactive it can show reduced or increased toxicity compared with the active parent compound (Bungaard, 1991).

A relatively benign toxicological profile with a racemic mixture may allow further development, but the problems encountered with thalidomide highlighted the fact that the toxicities of the S and R enantiomers may differ markedly. It is sometimes essential to determine the toxicology of the individual enantiomers even though the racemic mixture appears to cause little or no toxic changes, as the toxicity may be caused by an isomer-specific metabolite.

When the toxicity of chemical mixtures is under examination, the interpretation of data becomes more difficult: a mixture of two or more chemicals may result in a different qualitative or quantitative response relative to that predicted from the same but separate exposure to the mixture constituents. For example, the co-administration to rats of two hepatotoxic compounds, carbon tetrachloride and 1,2-dichlorobenzene, might be expected to show an additive effect on plasma alanine aminotransferase level (ALT), but the resultant dual exposure reduces the degree of liver injury and plasma ALT compared to the effects produced by 1,2-dichlorobenzene alone (Mumtaz *et al.*, 1993). Conversely, pretreatment of rats with retinol potentiates the hepatotoxic effects and change of associated plasma ALT with carbon tetrachloride (ElSisi *et al.*, 1993).

Polypharmacy may complicate toxic changes, and adverse effects may also be associated with 'inactive' ingredients in drug formulations (Golightly *et al.*, 1988). Some vehicles used for compound delivery markedly affect toxic changes, e.g. oily vehicles (Condie *et al.*, 1986).

Figure 1.1 indicates the central roles that blood and its filtration product, urine, play in the metabolism of xenobiotics. In toxicology studies, a variety of biochemical measurements can be used to evaluate a broad range of physiological and metabolic functions, identifying possible target organs, measuring impaired organ function, assessing the persistence and severity of tissue injury. Several of the common tests are interrelated and when combined these tests can provide better information from pattern recognition, e.g. urea and creatinine for glomerular function and several plasma enzymes for hepatotoxicity.

The findings obtained from clinical chemistry tests in animal toxicity studies have particular relevance for human studies where the same tests can be performed. The choice of tests and their applications are discussed in subsequent chapters, and some of the regulatory requirements for such studies are discussed in Chapter 2. The term 'biomarkers' is being used increasingly to describe several of the tests such as plasma enzymes which have been used for many years to detect toxicity, in addition to encompassing relatively new indices, e.g. plasma and urine profiles obtained with nuclear magnetic resonance spectroscopy (NMR) (Grandjean *et al.*, 1994; Timbrell *et al.*, 1994).

1.1 Predictive Values of Toxicology Studies

The prediction of toxic risk can be defined as the probability that an adverse event may occur or result from a particular challenge. However it must be remembered if the incidence rate of an adverse reaction is less than 0.01%, the adverse event which is apparent when a drug is administered to one million people, may be undetected in toxicology studies where the total number of animals may be less than one thousand. Animal toxicity studies may sometimes lead to erroneous elimination of promising drugs for man, and occasionally fail to predict toxicity

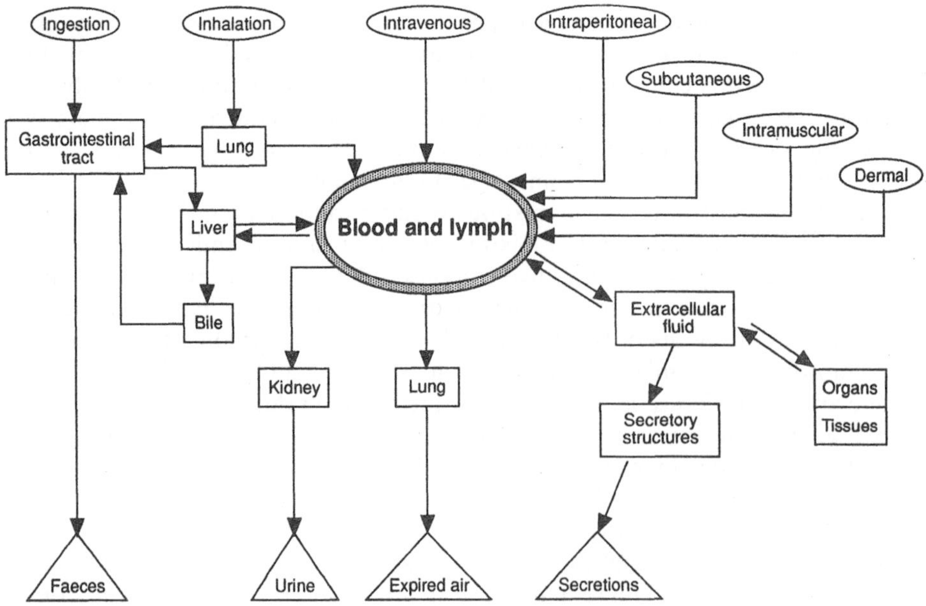

Figure 1.1 Ellipses indicate routes of administration: triangles indicate routes for excretion and secretion

which may occur in man. In a retrospective study of 45 new drugs, Fletcher (1978) showed that some adverse effects occurred exclusively in man and these were not detected in preclinical animal studies. Conversely, the two most commonly observed effects of hepatic and renal toxicity in animal studies were seen less frequently in man; these findings may reflect lower dosages, shorter periods of exposure and careful clinical monitoring. When animal toxicology studies have failed to detect adverse effects and predict toxicity in man, further demands are made by the public to improve the testing procedures and to re-examine the regulatory guidelines.

Some investigators choose to use the tools of test specificity and sensitivity to analyse the data from toxicity studies as defined in Table 1.1. Both specificity and sensitivity are desirable but are not always achievable. In retrospective analyses of 25 diverse anti-cancer agents, Schein and colleagues (1970, 1973) determined the predictive values for hepatoxicity in man using the formulae in Table 1.1 and data from dog and monkey studies. The measurements of aspartate aminotransferase and alkaline phosphatase in the dog and bromsulpthalein (BSP) clearance in the monkey gave the highest percentages of true positive and lowest percentages of false negative predictions: the BSP clearance tests over-predicted hepatotoxicity and did not add to the information obtained by the two enzymes (Thomann *et al.*, 1981).

The frequency of abnormal findings will depend upon the type and range of compounds being examined. In a survey of the toxicological profiles of 50 compounds – 38 pharmaceuticals and 12 agricultural or industrial chemicals – in rat, dog and monkey studies, Heywood (1981) found more than 30% of all compounds affected blood chemistry measurements, with over 18% affecting hepatic function and over 7% affecting kidney function tests.

Table 1.1 Definitions of test specificity and sensitivity

Sensitivity of an assay is the fraction of those with a specific disease that the assay correctly predicts.
Specificity is the fraction of those without the disease that the assay correctly predicts.

Where
TP = True positive, number of affected individuals correctly classified by the test
FP = False positive, number of non-affected individuals misclassified by the test
FN = False negative, number of affected individuals misclassified by the test
TN = True negative, number of non-affected individuals correctly classified by the test

Sensitivity is then expressed by $\frac{TP}{TP + FN} \times 100$

and specificity is expressed by $\frac{TN}{FP + TN} \times 100$

The predictive value of a positive test is $\frac{TP}{TP + FP} \times 100$

and the predictive value of negative test is $\frac{TN}{TN + FN} \times 100$

(Griner *et al.*, 1981).

Laboratory data produced for a study may be voluminous, particularly from a rodent study where more than 5000 clinical chemistry values may be obtained. Statistical analysis of these data can result in small but statistically significant differences between control and treatment groups which can be difficult to assess in many rodent studies (Carakostas and Banerjee, 1990). Glocklin (1983) suggested that: 'Frequently, statistically significant group differences in hematology or serum chemistry have no relevance unless certain of these changes correlate directly with each other or with some pathological lesion.' This assumes that the statistical methods chosen to analyse the data are appropriate but it fails to recognize that plasma or urine biochemistry may be a more sensitive indicator of toxicity, or that the timing of sampling in relation to exposure and necropsy is important. For example, the timing of the urine collections is a critical factor in detecting renal injury. Tubular necrosis is an event more likely to occur early in the study and therefore urine samples are best taken during this early period, whereas renal papillary necrosis is more likely to occur in the latter parts of a study (Heywood, 1981).

Relatively small changes are the rule rather than the exception and these small group changes may be indicative of early toxicity, even when they are within the accepted reference (or normal) ranges. In studies with larger laboratory animals, it is possible to detect trends due to treatment by establishing baseline data prior to the dosing period. The total variance for a single measurement may be expressed as the sum of analytical variance + biological variance + methodological (or study) variance + the interactions of these three components. In later chapters, we discuss some of these variables and how to approach the statistical analysis of the data.

Few if any of the common laboratory tests are intrinsically positive or negative, with most tests requiring some decision as to where to select the cut-off points for normality and abnormality. Effective interpretation involves considerably more than relying on probability values, it requires knowledge and experience – it is a deductive process. Clinical pathology measurements should always be interpreted in conjunction with the other study data obtained from histopathology, experimental and clinical observations. A simple example is plasma or urine osmolality where water intake, urine output and body mass should be collectively considered for the correct interpretation of the test.

1.2 Analytical Variables

Most laboratories use analysers and reagents designed for use with human samples and therefore the methods need careful evaluation before using with different species. Sometimes it is necessary to use methodology that can truly distinguish effects between control and treatment groups while accepting that limitations on accuracy exist, e.g. in endocrine assays where there is a lack of species-specific standards for several tests. The complex multi-test clinical chemistry analysers linked to computer systems are sometimes believed to be entirely faultless, but all systems must be thoroughly validated. The resulting analytical data must be reliable and the laboratory should operate internal quality control procedures and participate in external quality assurance (control) schemes to ensure consistently high performance standards. Adherence to the principles of Good Laboratory Practice (GLP) is essential for those laboratories that perform regulatory studies: these principles cover standard operating procedures, record systems, quality assurance, equipment etc. (Paget, 1979).

When applied to analytical procedures, the terms sensitivity and specificity have different connotations to those previously described for diagnostic procedures and they should not be confused. Analytical specificity relates to accuracy and refers to the ability to determine exclusively the analyte it claims to measure without being affected by other related substances. The potential for interference effects of a compound other than the analyte in question or the effect of a group of compounds on the accuracy of measurement of that analyte must be considered (see Chapter 3).

Analytical sensitivity is defined as the slope of the calibration curve and the ability of an analytical procedure to produce a signal for a defined change of unit. The limit of detection is defined as the smallest concentration or quantity of an analyte that can be detected with reasonable certainty for a given analytical procedure.

For each method, the laboratory should establish data which reflect the relative precision (or imprecision). Within-run (or batch) precision is the variability found when the same material is analysed repetitively in the same analytical run, or alternatively when duplicate analyses are made within a run. This can be extended to within-day precision and then further to day-to-day or between-day (or batch) precision where the variability found is for the same material analysed on different days. With modern instrumentation and commercially available high-quality reagents the analytical variance has been reduced, so that intra-batch coefficient of variation values of less than 5% are achievable for most of the common tests. It is usually a little higher for immunoassays and manual tests. Thus as a general rule, analytical variance is a smaller component than the biological variance in the total variance sum. Wherever possible, methodological changes during studies should be avoided.

The number of tests (or parameters) determined in a study is governed by the sample size and the analytical methodology available, although modern analysers generally require small sample volumes for most of the common tests (generally less than 40 μl per test). It is sometimes suggested that the choice of biochemical tests in animal studies has been governed by investigators more familiar with human medicine, and therefore this has led to the inclusion of tests which are inappropriate for the animal species selected. Such statements fail to recognize that several of the tests were originally evaluated in animal studies before their introduction to human medicine, and perhaps we need to increase our efforts in developing new tests with improved diagnostic accuracy. Knowledge of the toxicity of similar compounds, the structure or proposed use of a compound may suggest the inclusion of a particular test, e.g. pseudo-cholinesterase measurements with organophosphate compounds. The choice of tests is discussed in subsequent chapters.

1.3 Use of Animals

The three principles of reduction, refinement and replacement for the use of animals must be of concern to all investigators (Mann *et al.*, 1991). Whilst there is a growing public pressure to use alternatives such as *in vitro* studies to effect a reduction of the number of animals used, this still appears to be a long-term rather than a short-term goal in most applications. The challenge for clinical chemists and toxicologists is to extract more valuable information from the studies we perform in order that we improve risk:benefit judgements. The physiological effects of the removal of blood from smaller laboratory animals and the possible effects on test values must be considered (McGuill and Rowan, 1989; Evans, 1994; Chapter 3).

1.4 Biological and Chemical Safety in the Laboratory

Biological samples including blood and urine may present health risks to the personnel collecting and analysing the samples (Truchaud *et al.*, 1994). These risks are increased by exposure to laboratory chemicals and test compounds where the toxicological risks are unknown; Ballantyne (1993) has reviewed some of these additional hazards associated with work in toxicology laboratories. In general, precautions advocated for handling human samples can be equally applied to animal samples, and extended further where the risks appear to be greater, e.g. working with samples containing carcinogens, highly toxic compounds or from immunocompromised animals. Additional consideration must be given where staff might be exposed to potential mutagenic materials. Whilst the use of automated analysers reduces risk of exposure, problems can occur in the simple processes of sample collection and separation procedures, and with *in vitro* studies. Closed collection tube systems are currently designed for collecting larger blood samples than those usually obtained with smaller laboratory animals. Simple procedures such as disinfecting or autoclaving biological waste materials, and disinfection of laboratory analysers prior to selective maintenance procedures should be routine tasks.

1.5 Summary

Biochemical measurements can help in:

- Identifying target organ toxicity.
- Confirmation of other observations, particularly changes found by histopathological examination.
- Defining NOEL or NOAEL levels.
- Compound selection – screening of analogues.
- Elucidation of toxic mechanisms.
- Elucidation of interspecies differences in toxicity studies.

Clinical pathology must meet the future challenges and broaden its practice. We need to extend the repertoire of tests not for the sake of the exercise but to maximize the predictive values of the tests we offer for human and animal safety. We need to seek and provide further evidence for the correlation of biochemical changes with histological findings, and to gain a better understanding of the small but statistically significant changes that we detect. Other challenges include the need to support studies designed to reduce preanalytical variables, and to become more involved with nutritional studies where concern is currently focused on reducing obesity in the aged rat with little reference to alterations of biochemical values. Tietz (1994) has drawn attention to the need to return to the concept of accuracy in laboratory tests in human medicine, and this remains a constant challenge for the laboratories involved in animal studies where the lack of species-specific standards and methods remains.

References

BALLANTYNE, B. (1993) Hazards in the toxicology laboratory. In Ballantyne, B., Marre, T., & Turner, P., *General and Applied Toxicology*, pp. 359–67, Basingstoke: Macmillan.

BUNGAARD, H. (1991) Novel approaches to prodrug design. *Drugs of the Future*, **16**, 443–58.

CARAKOSTAS, M. C. & BANERJEE, A. K. (1990) Interpreting rodent clinical laboratory data in safety assessment studies: biological and analytical components of variation. *Fundamental and Applied Toxicology*, **15**, 744–53.

CONDIE, L. W., LAURIE, R. D., MILLS, T., ROBINSON, M. & BERCZ, J. P. (1986) Effect of gavage vehicle on hepatotoxicity of carbon tetrachloride in CD-1 mice: corn oil versus Tween aqueous emulsion. *Fundamental and Applied Toxicology*, **7**, 199–206.

ELSISI, A. E. D., HALL, P., SIM, W. W., EARNEST, D. L. & SIPES, I. G. (1993)

Characterisation of vitamin A potentiation of carbon tetrachloride-induced liver injury. *Toxicology and Applied Pharmacology*, **119**, 280–8.

EVANS, G. O. (1994) Removal of blood from laboratory mammals and birds. *Laboratory Animals*, **28**, 178–9.

FLETCHER, A. P. (1978) Drug safety tests and subsequent clinical experience. *Journal of the Royal Society of Medicine*, **71**, 693–6.

GLOCKLIN, V. C. (1983) The role of data organisation in the evaluation of toxicology studies in drug application. *Drug Information Journal*, 139–51.

GOLIGHTLY, L. K., SMOLINSKE, S. S., BENNETT, M. L., SUTHERLAND, E. W. & RUMACK, B. H. (1988) Pharmaceutical excipients. Adverse effects associated with 'inactive' ingredients in drug products. *Medical Toxicology*, **3**, 209–40.

GRANDJEAN, P., BROWN, S. S., REAVEY, P. & YOUNG, D. S. (eds) (1994) Biomarkers of chemical exposure. The proceedings of the Arnold O. Beckman/IFCC European Conference on Environmental Toxicology. *Clinical Chemistry*, **40**, 1359–476.

GRINER, P. F., MAYEWSKI, R. J., MUSHLIN, A. I. & GREENLAND. P. (1981) Selection and interpretation of diagnostic tests and procedures. *Annals of Internal Medicine*, **94**, 553–600.

HEYWOOD, R. (1981) Target organ toxicity. *Toxicology Letters*, **8**, 349–58.

JAMES, R. W. (1993) The relevance of clinical pathology to toxicology studies. *Comparative Haematology International*, **13**, 190–5.

MANN, M. D., CROUSE, D. A. & PRENTICE, E. D. (1991) Appropriate animal numbers in biomedical research in light of animal welfare considerations. *Laboratory Animal Science*, **41**, 6–14.

MCGUILL, M. W. & ROWAN, A. N. (1989) Biologic effects of blood loss: implications for sampling volumes and techniques. *ILAR News*, **31**, 5–18.

MIYA, T. S., GIBSON, J. E., HOOK, J. B. & MCCLELLAN, R. O. (1988) Contemporary issues in Toxicology. Preparing for the twenty-first century: Report of the Tox-90's commission. *Toxicology and Applied Pharmacology*, **96**, 1–6.

MUMTAZ, M. M., SIPES, I. G., CLEWELL, H. J. & YANG, R. S. H. (1993) Risk assessment of chemical; mixtures; biologic and toxicologic issues. *Fundamental and Applied Pharmacology*, **21**, 258–69.

PAGET, G. E. (1979) *Good Laboratory Practice*. Lancaster: MTP Press Ltd.

SCHEIN, P. & ANDERSON, T. (1973) The efficacy of animal studies in predicting clinical toxicity of cancer chemotherapeutic drugs. *International Journal of Clinical Pharmacology*, **8**, 228–38.

SCHEIN, P. S., DAVIS, R. D., CARTER, S., NEWMAN, J., SCHEIN, D. R. & RALL, D. P. (1970) The evaluation of anticancer drugs in dogs and monkeys for the prediction of qualitative toxicities in man. *Clinical Pharmacology and Therapeutics*, **14**, 3–40.

THOMANN, P., ACKERMANN, H. R. & ZIEL, R. (1981) Standard animal models of hepatotoxicity–species differences and relevance for man. In Davis, M., Tredger, J. M. & Williams, R. (eds), *Drug Reactions and the Liver*, pp. 321–37, London: Pitman Medical.

TIETZ, N. W. (1994) Accuracy in clinical chemistry – does anybody care? *Clinical Chemistry*, **40**, 859–61.

TIMBRELL, J. (1991) *Principles of Biochemical Toxicology*. London: Taylor & Francis.

TIMBRELL, J. A., DRAPER, R. & WATERFIELD, C. J. (1994) Biomarkers in Toxicology. New uses for old molecules? *Toxicology and Ecotoxicology News*, **1**, 4–14.

TRUCHAUD, A., SCHNIPELSKY, P., PARDUE, H. L., PLACE, J. & OZAWA, K. (1994) Increasing the biosafety of analytical systems in the clinical laboratory. *Clinica Chimica Acta*, **226**, S5–13.

ZBINDEN, G. (1993) The concept of multispecies testing in industrial toxicology. *Regulatory Toxicology and Pharmacology*, **17**, 85–94.

2

Study Design and Regulatory Requirements

D. T. DAVIES

For scientists involved in animal testing, it is necessary to design studies that generate the required safety assessment data and also satisfy the requirements of the various regulatory authorities which govern pharmaceuticals, agrochemicals and industrial chemicals. With a few notable exceptions, most regulatory authorities simply state the need for appropriate clinical pathology testing to be included without specifying exact requirements. Where individual tests are specified, a few may be considered inappropriate and scientifically unsound for use in animals. Several organizations (FDA, 1982; OECD, 1982; EPA, 1984; Malya *et al.*, 1988; MHW, 1991) have suggested different test profiles to be performed in regulatory studies, and the definition and acceptance of such profiles by regulatory and professional bodies is an ongoing process. Two notable initiatives, one national and a second international, to improve safety assessment and standardize regulatory testing guidelines, have recently been undertaken. Alongside these initiatives are the wider attempts to achieve international harmonization of other aspects of toxicological studies.

In the United Kingdom, the Organization for Economic Cooperation and Development (OECD) Shadow Toxicology Group made recommendations (Stonard, 1992) which will shortly be embodied in the Clinical Chemistry Sections of the OECD guidelines. The second initiative, seeking a universal agreement on clinical pathology testing, was undertaken when 10 national scientific organizations[1] formed a joint committee for the International Harmonisation of Clinical Pathology Testing (IHCPT) in order to provide expert recommendations for clinical pathology testing of laboratory animal species used in regulatory toxicity and safety studies. The recommendations of this committee (Weingand, 1994; Weingand *et al.*, 1996) and the salient points of these and the OECD recommendations, are discussed in this chapter.

2.1 Blood Sampling

Most regulations clearly state the need for four groups of animals (one control and a low, medium and high dose group) to be used in a safety assessment study and the number of animals per group are also usually defined. For sub-acute rodent studies, 10 males and 10 females per group are appropriate with double this number for chronic studies. For non-rodent studies (usually with dogs), four animals of each sex per group are usually sufficient for most regulatory studies. There are no clear guidelines as to the number of animals that should be bled at each time-point but, conventionally, all non-rodents should be bled while 20 rodents (10 males and 10 females) per group is considered appropriate. Similarly, the frequency of blood sampling is left entirely to the investigator, toxicologist or study director. As a minimum, blood samples should be collected at study termination. For mice, where limitation of blood volume makes interim blood sampling impractical, samples can only reliably be collected at necropsy. Multiple blood sampling from the jugular or cephalic veins is clearly not a problem in the larger non-rodents while, in the rat, several acceptable sites of interim blood sampling are described in Chapter 3.

Blood chemistry and haematology results can be influenced by the conditions under which blood is collected, for example feeding or fasting, the anaesthetic used, whether the animals are bled randomly and the site of blood collection. For these reasons, it is advisable that the blood should be collected in the same manner at each sampling time in order to minimize variation and that the volume of blood collected should be restricted to 1.5 ml or 1% of body mass, whichever is the smallest. The OECD Shadow Toxicology Group did not see a need for fasting animals prior to the collection of blood for clinical chemistry measurements in a routine screening approach. Rather, it was felt that the measurement of non-fasting glucose was a more sensitive measure than fasting glucose. However, the importance of fasting for a more in-depth investigation of a particular effect on carbohydrate or lipid metabolism was recognized.

Pre-study clinical pathology testing is useful for general health screening of non-rodent species considered for placement in toxicity and safety studies. Pre-study clinical pathology testing of rodents is not generally necessary because of the large number of animals per group and the general homogeneity of the laboratory animal population. Additionally, pre-study blood sampling of young rodents is limited by their size and may have adverse effects on animal health. Furthermore, evaluation of test compound effects in rodents by comparison with the clinical pathology results obtained pre-study is not considered appropriate since many tests are age dependent and change significantly in rapidly growing young rats.

During the study, the frequency and timing of clinical pathology testing are dependent upon study duration, study objectives, the biological activity of the test material and the species tested. For repeated-dose studies in non-rodent species, clinical pathology testing is recommended at study termination and at least once at an earlier interval because of the small number of animals per group and inter-animal variability. For 13-week studies, it is suggested that blood samples be taken at approximately 4 and 13 weeks while, for chronic studies, additional blood samples should be considered at 26 and 52 weeks. For studies of 2- to 6-week duration, testing is also appropriate within 7 days of initiation of dosing. For rodent repeated-dose studies, clinical pathology testing is essential at study termination.

Interim blood sampling of rats may not be necessary provided that short duration studies have been done using dose levels that are not substantially lower than the longer duration studies. Study needs may, however, require interim blood sample analysis and the OECD Shadow Toxicology Group firmly believe that blood samples taken during the livephase of a sub-chronic study, when pathological evaluation of tissue is not available, exploit the strengths of clinical pathology and provide additional screening information of value. A blood-sampling regimen, similar to that described for non-rodents, can be adopted provided that the sampling does not compromise the health of the animals.

This view is further supported by Davies (1992) who demonstrated that early interim blood samples can provide valuable information on the effects of drugs on clinical pathology. In the experiment described, blood samples taken at the end of a 3-month dog study had normal plasma alkaline phosphatase (ALP) and aspartate aminotransferase (AST) activities but there was a slight increase in alanine aminotransferase (ALT) activity; morphologically the liver was normal. Early blood samples taken during weeks 1 and 2 exhibited elevated ALP and ALT activities and the mean ALT value at week 2 showed more than a 20-fold increase compared to the pre-study results. These elevated values persisted for at least 5 weeks but, despite continued dosing, the liver had virtually recovered by the end of the 3-month dosing period. If terminal samples only had been taken in this study, valuable information would have been missed. Similar examples can be cited for haematological changes following the administration of a cytotoxic drug that has an effect on bone marrow. With such compounds, the haematological profile goes through a cycle of reduced white and red cell counts followed by a recovery to pre-dose values.

In common with most species, rodents as they age become progressively more susceptible to naturally occurring diseases that can modify clinical pathology results and thus obscure meaningful interpretation. Quantitative clinical chemistry and urinalysis should, therefore, be generally avoided in rodents during their second year of life. Clinical pathology evaluation in these animals, generally undertaken towards the end of a carcinogenicity study, should be confined to a haematological evaluation of blood smears taken from decedents and at study termination to aid in the interpretation and differentiation of haemopoietic neoplasia. For both rodents and non-rodents, clinical pathology testing should be performed on the same animals as examined for morphological pathology findings and it is also recommended that blood samples from recovery or withdrawal satellite groups should be analysed at study termination.

Although most modern analysers require only micro-volumes of plasma or serum to perform a clinical chemistry test (e.g. 2 μl of sample is needed for glucose determination on the Hitachi 747 autoanalyser), problems with blood collection may mean that insufficient sample is obtained for the analysis of all the designated tests. Some investigators suggest pooling samples for animals in the same group or cage and analysing the pooled samples. This approach is not recommended since it can lead to difficulties when interpreting the results. For example, pooling a grossly haemolysed sample with four non-haemolysed samples will markedly increase the pooled potassium value; however, there is no way of knowing how many samples were affected. When samples of reduced volume are received, it is preferable that the tests should be prioritized such that the important tests are analysed first for each individual. In extreme circumstances a further blood collection for specific animals should be considered. If an inadequate amount of blood is a regular occurrence, possibly because of large blood volumes required

Table 2.1 Pre-clinical blood chemistry tests cited in regulatory documents

Glucose*	Inorganic phosphate*
Urea (or urea nitrogen)*	Urate*
Creatinine*	Cholesterol*
Aspartate aminotransferase*	Lactate dehydrogenase*
Alanine aminotransferase*	Total protein*
Total bilirubin*	Albumin*
Total calcium*	Electrolytes*
Succinate dehydrogenase	Alkaline phosphatase
Gamma glutamyl transferase	Triglycerides
Ornithine carbamoyl transferase	Ornithine decarboxylase
Creatine kinase	Chloride
Protein electrophoresis	5′-nuceotidase
Cholinesterase	Various lipids
Acid–base balance	Various hormones

*denotes test referenced by Alder *et al.* (1981).

for certain analyses (e.g. measurement of plasma hormones), the investigator may wish to solve the problem by allocating additional animals for these analyses. Samples can be then taken from sub-groups rather than try and fail to collect sufficient samples from all animals.

2.2 Clinical Chemistry Tests

In a review of the pre-clinical requirements of 19 countries Alder *et al.* (1981) identified a total of 14 clinical chemistry tests that should be performed: not all of these tests were required by all of the countries surveyed. The requirements for other tests have also appeared in other documents since this time and the complete list, displayed in Table 2.1, is quite daunting.

Although all of these tabulated tests were at one time considered appropriate, more recently some have been criticized as being useful in human medicine but not necessarily useful in other species. For example, as illustrated in Chapter 5 the distribution of common enzymes varies between the different species such that an increase in plasma activity indicative of hepatotoxicity in one species may not be apparent in a second.

As discussed in Chapter 6 (Hepatotoxicity), because the renal clearance of bilirubin is very high in the rat the test is rarely sensitive enough to detect minor liver functional change in rodents. Similarly, while gamma glutamyl transferase is a useful marker of liver toxicity in the dog, an increase is rarely detected in rodents when exposed to a hepatotoxic compound. Other tests suggested by regulatory authorities and professional associations have also proved to be controversial.

Urate (uric acid) measurement is commonly used in human medicine for the diagnosis and treatment of gout, rheumatoid arthritis and other conditions that can alter purine metabolism. Urate is the end point of purine metabolism in humans, but most mammalian species are able to further metabolize urate to allantoin resulting in lower plasma urate values which have little diagnostic value in general toxicity studies. Lactate dehydrogenase activity is highly variable and lacks specificity as an indicator of major organ toxicity in animal species (Evans, 1991). The analytical requirements and testing performance for plasma ornithine decarboxylase (Carakostas, 1988) and to a lesser extent ornithine carbamoyl transferase (Carakostas *et al.*, 1986) make these tests impractical for use as routine screening tests in non-clinical toxicity and safety studies. Succinate dehydrogenase is rarely measured in toxicity studies and few data have been published on its utility. Similarly, there is currently very little scientific information available concerning the effects of toxicity on serum protein fractions in laboratory animal species to support meaningful interpretation of electrophoretic separation of serum proteins for many of the samples taken routinely during the study.

Some tests, although not appropriate for routine screening for adverse toxicological effects, have an interpretive value when used appropriately. The measurement of cholinesterase is particularly relevant to test substances which are carbamates, organophosphates or other chemicals which may predictably inhibit cholinesterase (see Chapter 11). Creatine kinase, although highly variable, is an excellent indicator of striated muscle damage in short-term duration studies where blood samples are collected at frequent intervals during the study. Similarly, the use of serum protein electrophoresis can be valuable if the clinical pathologist is trying to demonstrate an acute-phase response when an inflammation is clinically observed or suspected. Under these circumstances blood samples are usually taken at frequent intervals at times expected to coincide with the peak of the inflammatory response.

The 'regulatory check box' approach to testing is commercially important but it should always be supported by a consideration of the scientific validity of the chosen tests. It is not uncommon for a safety strategy to include a 'standard' set of reference tests that are included in every study. The appropriate selection of these tests should ensure that no adverse clinical pathology event remains undetected and, in the absence of any toxicity, no additional tests would be required to satisfactorily complete the safety evaluation package. Adopting this approach does not prevent the inclusion, in subsequent studies, of additional tests that can provide further useful information.

A tiered approach to regulatory studies seems highly desirable, with the selection of an initial 'core' list of tests and the use of additional tests where these are required to further characterize toxicity. Core tests should include general measurements of the common target organs, e.g. plasma urea and creatinine for renal injury and several plasma enzymes for hepatic injury. If target organ toxicity is observed in 'sighting or dose ranging' studies, then further tests can be added to the pivotal studies to characterize this toxicity, e.g. in a study where the kidney is the target organ and proximal tubular damage is suspected, several urinary enzymes can be measured and urine cytology can be considered (see Chapter 7, Nephrotoxicity).

The debate about which tests should form the 'core' profile for inclusion in all non-clinical toxicity and safety studies has exercised several committees over the years. Recent noteworthy recommendations are discussed below. As mentioned previously, two groups have been prominent in publishing recommendations for clinical pathology testing. In 1992, Stonard published the recommendations of the OECD UK Shadow Toxicology Group which formulated draft OECD Guideline (407/1994) for the testing of chemicals. Subsequently, the IHCPT Committee developed its own recommendations from an earlier version prepared jointly by the American Association of Clinical Chemistry and the American Association of Veterinary Clinical Pathology (Weingand *et al.*, 1992). Both documents emphasize that they define **minimum recommendations** and more tests may be required to fully characterize any observed effect. The recommended core tests to be included in all clinical chemistry profiles are displayed in Table 2.2.

All the tests recommended by the OECD Shadow Toxicology Group are also recommended by the IHCPT. The OECD Shadow Toxicology Group also suggested that additional enzyme measurements (of hepatic or other origin) may provide useful information under certain circumstances. Similarly, tests such as: calcium, inorganic phosphate, fasting triglycerides, fasting glucose, specific hormones, methaemoglobin and cholinesterase may also be performed but these must be identified on a case-by-case basis. The IHCPT recommended that some of these additional tests should be included as part of the core clinical chemistry profile. Further information on the merits and relevance of these tests for screening and toxicological evaluation of new compounds, are discussed at length in the relevant chapters of this book.

2.3 Urinalysis

When collecting urine for clinical chemistry evaluation every effort must be made to avoid contamination with food, drinking water and faeces. Urine samples should not be pooled and it is critical that the urine should be collected in the same manner for all animals, at least once during the study. Urine sample quality can be maintained by collecting samples into a cooled container or one containing a suitable preservative. For routine urine analysis an overnight collection (approximately 16 hours) into individual collecting vessels containing preservative is recommended.

The IHCPT recommend that the core tests should include an assessment of urine appearance (colour and turbidity), volume, specific gravity or osmolality, pH and either the quantitative or semi-quantitative determination of total protein and glucose. Microscopic urine sediment examination and urinary mineral and electrolyte excretion are not considered to be very useful as routine screening tests. These recommendations are similar to those of the OECD Shadow Toxicology Group with the proviso that they consider the 'dipstick' method of limited value, especially in rodents, since many low molecular weight proteins are less well detected. Additional tests, such as the measurements of urinary enzymes of renal origin, with the appropriate frequency of urine collection, may also be necessary for evaluating test compounds suspected of causing renal toxicity (Stonard, 1990; Chapter 7, Nephrotoxicity).

2.4 Good Laboratory Practice (GLP)

Following the identification of serious flaws in toxicological studies submitted to the Food and Drug Administration (FDA) of the United States of America, the FDA developed a code of practice designed, with periodic inspections by government agencies, to ensure the scientific validity of all studies submitted. This

Table 2.2 Recommendations for core clinical chemistry tests

OECD Shadow Toxicology Group	IHCPT
Non-fasting glucose	Glucose
Urea	Urea (blood urea nitrogen)
Creatinine	Creatinine
Total protein	Total protein
Albumin	Albumin
Sodium	Calculated globulin
Potassium	Total calcium
Total cholesterol	Sodium
Alanine aminotransferase	Potassium
Bile acids	Total cholesterol
Total bilirubin	*Hepatocellular tests*
	[Two from:]
	Alanine aminotransferase
	Aspartate aminotransferase
	Sorbitol dehydrogenase
	Glutamate dehydrogenase
	Total bile acids
	Hepatobiliary tests
	[Two from:]
	Alkaline phosphatase
	Gamma glutamyl transferase
	5′-nuceotidase
	Total bilirubin
	Total bile acids

Good Laboratory Practice (GLP) Regulation (FDA, 1979), together with OECD guidelines (OECD, 1982) were subsequently promulgated in most of the developed countries.

As for all other aspects of non-clinical toxicity and safety studies, all the clinical pathology analyses must be shown to be scientifically valid and must meet GLP standards. These GLP regulations cover aspects of personnel and management including training, general laboratory facilities, calibration and maintenance of equipment, performance of reagents and therefore assays, and characterization of other experimental variables. Great emphasis is placed on adequate documentation, a clear identification of 'raw data' and detailed description of any amendments made to the original results.

Appropriate statistical methods should be used to analyse clinical pathology data (Gad and Weil, 1989) and it must be stressed that concurrent control data are more appropriate than historical reference ranges for comparison with compound treatment groups. Regardless of the outcome of statistical analysis, scientific interpretation is paramount for the ultimate determination of compound effects. Statistical significance alone should not be used to infer toxicological or biological relevance of clinical pathology findings. Additionally, the absence of statistical significance should not preclude the possibility that compound treatment effects exist.

2.5 Conclusions

These recommendations provide a clear guide for the **minimum** clinical pathology that should be included in a toxicology safety evaluation study. Clinical pathology requirements in toxicity guidelines cannot be considered in isolation but must form part of a coherent approach to the testing strategy. The need for a certain degree of flexibility and judgement in what should be measured, over and above the stated minimum recommendations, is clearly of paramount importance in this approach. Strict adherence to a list of prescribed measurements, some of which are either not applicable or completely irrelevant, benefits neither those who provide the data nor those who receive the data. Both the OECD Shadow Toxicology Group and the IHCPT Committee hope that their recommendations will serve as the basis for international harmonization of clinical pathology testing guidelines developed by various government regulatory agencies and professional standards organizations. For the toxicologist or study director developing a study protocol for the safety assessment of a new compound they provide clear guidance of which tests should always be included.

Note

1 IHCPT participating organizations: American Association for Clinical Chemistry's Division of Animal Clinical Chemistry (DACC-United States); American Society for Veterinary Clinical Pathology (ASVCP-United States); Animal Clinical Chemistry Association (ACCA-United Kingdom); Arbeitsgruppe Klinische Chemie bei Laboratoriumstieren der Deutschen Gesellschaft für Klinische Chemie E.V. (AKCL-Germany); Association des Biologistes Cliniciens pour Animaux de Laboratoire (ABCAL-France); Association of Comparative Haematology (ACH-United Kingdom); Dutch Association for Comparative Haematology (DACH-Netherlands); Gruppo di Ematologia ed Ematochimica Applicate alla Tossicologia (GEET-Italy); International Society for Animal Clinical Biochemistry (ISACB-Israel, Canada, Sweden); Japanese Pharmaceutical Manufacturers' Association (JPMA-Japan).

References

ALDER, S., JANTON, C. & ZBINDEN, G. (1981) Preclinical safety requirements in 1980. Swiss Federal Institute of Technology and University of Zurich.

CARAKOSTAS, M. C. (1988) What is serum ornithine decarboxylase? *Clinical Chemistry*, **34**, 2606–7.

CARAKOSTAS, M. C., GOSSETT, K. A., CHURCH, G. E. & CLEGHORN, B. L. (1986) Evaluating toxin-induced hepatic injury in rats by laboratory results and discriminant analysis. *Veterinary Pathology*, **23**, 264–9.

DAVIES, D. T. (1992) Enzymology in preclinical safety evaluation. *Toxicologic Pathology*, **20**, 501–5.

EPA (1984) US Environmental Protection Agency. Pesticide assessment guidelines. National Technical Information Service, Springfield VA, USA EPA/540/9-84/014.

EVANS, G. O. (1991) Biochemical assessment of cardiac function and damage in animal species. *Journal of Applied Toxicology*, **11**, 15–21.

FDA (1979) US Food and Drug Administration, Nonclinical Laboratory Studies: Good Laboratory Practice Regulations. Federal Register. **43**, 59986–60025.

(1982) US Food and Drug Administration, Bureau of Foods, In Toxicological principles for the safety assessment of direct food additives and colour additives used in food. National Technical Information Service, Springfield VA, USA PB 83-170696.

GAD, S. C. & WEIL, C. S. (1989) Statistics for toxicologists. In Hayes, A. W. (ed.) *Principles and Methods of Toxicology*, 2nd Edn, pp. 435–83, New York: Raven Press.

MALYA, P. A. G., NACHBAUR, J., DOOLEY, J. F., BRAUR, J., GALTEAU, M. M., SIEST, G. & TRYDING, N. (1988) Drug interference tests on laboratory animals during toxicity studies. *Journal of Clinical Chemistry and Clinical Biochemistry*, **26**, 175–9.

MHW (1991) Ministry of Health and Welfare in Japan, 1990. Guidelines of Toxicity Studies of Drugs Manual, Editorial Supervision by New Drugs Division, Pharmaceutical Affairs Bureau, MHW, Yakuji Nippo Ltd, Tokyo.

OECD (1982) Organization of Economic Cooperation and Development's Principle of Good Laboratory Practice, OECD Guidelines for Testing of Chemicals. C(81)30 Final Annex 2.

STONARD, M. D. (1990) Assessment of renal function and damage in animal species. *Journal of Applied Toxicology*, **10**, 267–74.

(1992) Clinical Biochemistry Requirements in the Organization for Economic Cooperation and Development Guidelines – United Kingdom Recommendations. *Toxicologic Pathology*, **20**, 506–8.

WEINGAND, K. (1994) Recommendations of the Joint Scientific Committee for International Harmonization of Clinical Pathology Testing in Toxicity and Safety Studies. DACC Newsletter Autumn 1994.

WEINGAND, K., BLOOM, J., CARAKOSTAS, M., HALL, R., HELFRICH, M., LATIMER, K., LEVINE, B., NEPTUNE, D., REBAR, A., STITZEL, K. & TROUP, C. (1992) Clinical pathology testing recommendations for nonclinical toxicity and safety studies. *Toxicologic Pathology*, **20**, 539–43.

WEINGAND, K., DAVIES, D. T. *et al.* (1996) Recommendations of the Joint Scientific Committee for International Harmonization of Animal Clinical Pathology Testing in Toxicity and Safety Studies. *Fundamental and Applied Toxicology* (in press).

3

Preanalytical and Analytical Variables

J. ROBINSON & G. O. EVANS

It is essential to distinguish between responses produced by a test compound and the changes which occur in plasma and urine as normal biological variations, particularly where statistical differences are found for data in the absence of any other findings. For each study, we can consider several variables or experimental steps: these variables can be divided into (1) biological, (2) analytical and (3) pharmacological or physiological effects due to the test compound: these three components may interact in a synergistic or antagonistic manner. Sometimes these variables are expressed as:

1 Biological variance
2 Analytical variance
3 Methodological variance
4 Interactions of 1, 2 and 3

Sum = Total experimental variance

The methodological variance is introduced by the experimental procedures during the study. This variable is important when data are compared for the same test compound where the route of administration may differ or the study is repeated in another centre using slightly different experimental procedures.

The timing of blood or urine sampling in relation to the toxic insult is critically important. For example, following renal tubular damage or cardiac muscle injury, enzyme changes may occur rapidly in the early phase of a study, and values may have returned to normal or pretreatment levels at the times scheduled for sample collection (see Chapter 2 – Study Design and Regulatory Requirements).

In this chapter, we discuss these experimental variables in two separate sections as preanalytical and analytical factors: these two factors can affect both test and control group animals. The third section deals with analytical effects due to the presence of the test compound in the sample.

Section 3.1 Preanalytical Variables

J. ROBINSON & G. O. EVANS

Several preanalytical variables must be considered, and these include species, genetic influences, gender, age, environmental conditions, chronobiochemical changes (or biorhythms), nutrition, fluid balance and stress (Table 3.1.1). In addition, the procedures used for sample collection, separation and storage are also important.

3.1.1 Species, Strain, Age and Gender

Clearly there are major differences between species for reference clinical chemistry values in healthy laboratory animals (Mitruka and Rawnsley, 1977; Caisey and King, 1980; Loeb and Quimby, 1989; Matsuzawa *et al.*, 1993), and these differences are not necessarily related to size or relative organ weight (Garattini, 1981). Marked differences between species occur in enzymes (Chapter 5), hormones (Chapter 8) and lipids (Chapter 13). These interspecies differences for clinical chemistry tests must be recognized in the same way that we recognize the different metabolic response to xenobiotics in the various species.

Genetic differences may occur in healthy animals, e.g. glomerular filtration rates vary between inbred strains of rats (Hackbarth *et al.*, 1981). Some strains such as diabetic rats or the hyperlipidaemic Watanabe rabbit may be chosen specifically to test for toxic effects of therapeutic agents, and the biochemical profiles of these strains will differ from the strains used for general toxicology studies.

Differences related to age and gender of laboratory animals are often observed (Nachbaur *et al.*, 1977; Nakamura *et al.*, 1983; Uchiyama *et al.*, 1985; Wolford *et al.*, 1987). Plasma alkaline phosphatase is probably the best-known example of a parameter which is age dependent, with relatively high values during the early growth phase associated with osseous activity followed by lower values in mature animals. Plasma creatinine values also tend to increase with age reflecting changes in muscle mass and maturity, and later alterations in the glomerular filtration rate. In rats, proteinuria (albuminuria) increases with age with consequential changes of plasma proteins occurring at approximately 8 to 14 months of age. Apart from

Table 3.1.1 Preanalytical variables

Animals
Species
Strain
Gender
Age
Environment
Nutritional status
Fluid balance
Chronobiochemistry
Sampling procedures
Time of sampling
Sample volume
Site of sampling
Agents used for anaesthesia or euthanasia
Frequency of sampling
Anticoagulant
Sample quality
Sample storage

the obvious sex differences observed for the reproductive hormones, other plasma and urine parameters differ. Male rats excrete the major urinary protein, alpha$_{2u}$ globulin, which is not detectable in female rat urine.

3.1.2 Environment

Several environmental factors including caging density, lighting, room temperature, relative humidity, cage bedding, cleaning procedures etc. have been shown to affect animals (Fouts, 1976; Riley, 1981; Vadiei *et al.*, 1990; Whary *et al.*, 1993). Modern animal accommodation is now designed to minimize but does not eliminate these effects. With today's breeding and animal care procedures, parasitic or viral infections are less likely to occur, but these infections when present may affect some measurements, e.g. plasma proteins.

Transportation within and between animal accommodation can affect some biochemical tests (Bean-Knudsen and Wagner, 1987; Kuhn and Hardegg, 1988; Drozdowicz *et al.*, 1990; Garnier *et al.*, 1990; Toth and January, 1990; Tuli *et al.*, 1995), and these are often the tests used as measures of stress, e.g. plasma corticosterone and catecholamines (Vogel, 1987). It is advisable not to collect samples from animals immediately after delivery, except where it is essential for health checks; animals should be allowed to acclimatize to their new environment before samples are taken.

3.1.3 Nutrition and Fluid Balance

While in recent years, attention has been drawn to the effects of differing diets on longevity and tumour incidence rates, long-term carcinogenicity studies, diet

Table 3.1.2 Mean percentage changes observed in rats fasted overnight compared to non-fasted controls at 4 and 13 weeks of age (Jenkins and Robinson, 1975)

	Mean percentage changes			
	4 weeks		13 weeks	
Parameters	Male	Female	Male	Female
Body mass	−11	−9	−2	−3
Absolute liver mass	−34	−33	−20	−20
Plasma biochemistry				
Urea	−19	−29	−34	−41
Creatinine	+20	+14	+4	+13
Total cholesterol	−22	−11	+5	−6
Triglyceride	−44	−39	−40	−30
Aspartate aminotransferase	+11	+16	+11	+12
Alanine aminotransferase	−13	−5	0	−15

or feeding regimes may also cause a marked effect on biochemical values, particularly in small rodents. Food restriction for 16 to 24 h in rats may produce changes in the composition of plasma enzymes, urea, creatinine and glucose (Table 3.1.2; Jenkins and Robinson, 1975; Apostolou *et al.*, 1976; Kast and Nishikawa, 1981; Matsuzawa and Sakazume, 1994). The changes associated with overnight food deprivation in smaller rodents, which include altered organ weights, suggest that this is undesirable although several regulatory documents list this as a requirement (Matsuzawa and Sakazume, 1994; see Chapter 2). Post-prandial elevations of plasma urea and creatinine have been reported in dogs (Street *et al.*, 1968; Evans, 1987).

Longer-term studies in rats (Schwartz *et al.*, 1973; Oishi *et al.*, 1979; Levin *et al.*, 1993), mice (Dancla and Nachbaur, 1981) and cynomolgus monkeys (Yoshida *et al.*, 1994) have also shown the effects of dietary restriction on plasma biochemical tests. Studies in one laboratory (J. Robinson, unpublished data) have shown that some plasma values changed when rats were fed a semi-purified diet versus a diet consisting of refined oats, soya proteins and fish meal: in this study, plasma alanine aminotransferase values were markedly lower in animals fed the semi-purified diet. Changes in the composition of dietary fat or the use of corn oil as a vehicle for test compounds also produce effects on plasma chemistry measurements (Meijer *et al.*, 1987). Mineral imbalances in the diets may cause nephrocalcinosis in some species with associated changes in plasma and urine biochemistry (Stonard *et al.*, 1984; Bertani *et al.*, 1989; Meyer *et al.*, 1989).

The effects of dehydration on plasma volume and thus biochemical measurements may be obvious with high values for plasma protein and packed cell volume. High fluid intake with consequential perturbations of fluid balance also occur (Boemke *et al.*, 1990). Simple observations of water intake together with urinary output are important when examining this preanalytical variable.

Treatment prior to sequential blood sampling in toxicology studies should be similar in respect to food and water intake. The biochemical changes which occur with alterations of nutritionary and fluid balance can be associated with alterations in the absorption or uptake of the test compound, changes in metabolism rates or detoxification mechanisms, modification of renal clearance or changes in the competitive binding of proteins with the test compound. Toxicity may affect both nutrition and fluid balance and exaggerate differences between dosed or treatment groups.

3.1.4 Chronobiochemical Effects

Periodic or cyclic variations and their effects on toxicology and pharmacology are being increasingly recognized (Scheving *et al.*, 1993), and periodic changes occur for several analytes, particularly hormones. Examples of short-term changes include diurnal variations for cortisol in dogs (Orth *et al.*, 1988) and glucose in mice. Urinary volumes in mice and rats vary during any 24 h period (Myers and MacKenzie, 1985), and can be altered by fluid administration. Longer-term cycles are observed for reproductive hormones (see Chapter 8; Nigi and Torii, 1991; Torii and Nigi, 1994). Changes of accommodation, feeding patterns (Maejima and Nagase, 1991), light–dark cycles or other experimental procedures can alter these periodic fluctuations in plasma and urinary measurements.

3.1.5 Blood Collection Procedure

In all procedures when collecting blood from laboratory animals, the sample volume, site, anaesthetic agent, stress and anticoagulant must be considered, together with the relationship between the sample volume and the total blood volume. All procedures are governed by local and national recommendations and regulatory guidance for animal welfare.

As a general rule, the blood volume is approximately 6–7% of the body volume, i.e. 60–70 ml/kg body weight, and the volumes from conscious animals will be less than at necropsy; this is particularly important for smaller animals. The collection of 0.5 ml of blood from a 20 g mouse represents a 20% loss of blood volume for that animal. Repetitive sampling or excessive blood withdrawal may cause anaemia, and prejudice the outcome of a study. McGuill and Rowan (1989; see also Evans, 1994) suggested investigators limit blood sampling to 15% of the total blood volume for a single sampling and 7.5% as a weekly limit for repetitive blood sampling.

The choice of sampling site depends on the volumes of blood required for analysis, and whether the sampling time is during or at the end of a study when larger samples can usually be taken. The sites for blood collection used for different species vary, and sites are sometimes preferentially chosen on the basis of local experience with a technique, in the absence of data which suggest that one site is better than another. Some blood collection procedures require a greater degree of operator skill, and Fowler (1982) has reported differences in sample quality associated with individual operators. Table 3.1.3 shows some of the common sites used for blood collection from several species.

Anaesthetics should be chosen on the basis of causing minimal stress to the animal, minimal interference effects on the analyte, and minimizing the hazards

Table 3.1.3 Species and common sites for blood collection

Species	Sites of collection
Mouse	heart, vena cava, tail vein
Rat	heart, vena cava, aorta, tail vein, retro-orbital plexus, jugular vein
Hamster	heart, jugular vein
Guinea pig	heart
Rabbit	heart, ear vein or artery, jugular vein
Dog	heart, cephalic vein, jugular vein, saphenous vein
Primate	femoral artery, jugular vein, cephalic vein, heart

to the operator through accidental injection, chemical toxicity of the anaesthetic agents or injuries caused by the animal's behaviour during the procedure. Anaesthetic agents used for rodents include halothane, ether, barbiturate, methoxyfluorane and carbon dioxide. Frequently repeated anaesthesia can affect the analyte values. In the experience of one of the authors (J. Robinson, unpublished data), halothane with a closed system or a mixture of carbon dioxide and oxygen are suitable anaesthetics for rodents.

The effects of different procedures on several biochemical measurements have been reported in rats (Neptun *et al.*, 1985; Suber and Kodell, 1985), and dogs (Jensen *et al.*, 1994). The use of restraining procedures may affect biochemical values in small and large laboratory animals (Gartner *et al.*, 1980; Swaim *et al.*, 1985; Davy *et al.*, 1987; Landi and Kissinger, 1994). In general, the collection of blood from dogs and most primates does not cause particular problems in terms of the sample volume required for clinical chemistry (Mitruka and Rawnsley, 1977; Fowler, 1982).

Combinations of different methods of blood collection within the same study for interim and terminal sampling points may confound interpretation of data and this should be avoided it at all possible. When dealing with a study where there are a large number of animals in several treatment groups, blood collections should be randomized to avoid introducing a bias, which may occur if groups of animals are bled sequentially. The collection order should also be randomized if several individuals are responsible for collecting samples. Pooling of blood samples from several animals should be avoided as this procedure hides intra-animal variation. The possible use of a wide range of procedures emphasizes the caution which is needed when comparing study data with reference ranges or published data, and the need for contemporary controls.

3.1.6 Anticoagulants

Lithium heparinate or heparin are suitable anticoagulants for most plasma measurements. Inappropriate use of anticoagulants, and incorrect proportions of anticoagulant tó blood volumes may also cause errors. Samples collected with sequestrene (EDTA) or sodium citrate used for haematological investigations are not suitable for several electrolyte and enzyme measurements; the inappropriate use of potassium EDTA can be recognized by low calcium values due to chelation and high potassium values. Investigators should check with local laboratories the correct anticoagulant required for particular tests.

When collecting blood, it is important to separate the plasma or serum as soon as possible; this reduces the effects of glycolysis which result in reduced glucose levels, and increased lactate, inorganic phosphate and potassium values. Sodium fluoride or fluoride/oxalate anticoagulants may be used to minimize the effects of glycolysis, but these anticoagulants interfere with other measurements including enzymes and electrolytes.

For several enzyme measurements, it is preferable to use plasma rather than serum because of the relatively high erythrocytic concentrations of those enzymes, or other enzymes which may interfere with their measurements (Friedel and Mattenheimer, 1970; Korsrud and Trick, 1973). The presence of platelets in plasma or serum as a result of inadequate separation procedures may cause erroneous results for potassium and lactate dehydrogenase (Evans, 1985; Reimann *et al.*, 1989). Platelet counts may also be a useful indicator of sample quality, where low counts may indicate poor quality in the absence of any effect due to the test compound.

3.1.7 Haemolysis, Lipaemia and Storage

Observations of haemolysis and lipaemia should always be recorded. Forcing blood rapidly through a narrow gauge needle is a common cause of haemolysis in the absence of an effect caused by the test compound. The interference effects of haemolysis on clinical chemistry analytes have been reported in several species (Chin *et al.*, 1979; Sonntag, 1986; O'Neill and Feldman, 1989; Reimers *et al.*, 1991; Jacobs *et al.*, 1992).

Interference with some measurements may be caused by turbidity associated with hyperlipaemia, or the interference may be the result of alterations to the plasma volume (McGowan *et al.*, 1984). The turbidity due to lipaemia may be reduced by ultracentrifugation or precipitation techniques (Thompson and Kunze, 1984), but these procedures may disguise significant changes of analyte concentrations.

The stability of several components of plasma or serum may alter depending on both the storage temperature and period (Hirata *et al.*, 1979; Falk *et al.*, 1981; Szenci *et al.*, 1991). Cyclical freezing and thawing of samples should be avoided as this denatures proteins. With small volumes, analytical problems may occur because of evaporation of the samples.

3.1.8 Urine Collections

In general urine collections should be made for fixed time periods (often 17 h or overnight) using well-designed metabolism cages. For some analytes such as urinary enzymes the timing of collection period may produce differing results. The design of devices for separating faeces and urines in metabolism cages is a critical factor in the reduction of faecal contamination of the urine. Bacterial growth, often indicated by highly alkaline pH values, may occur if analysis is delayed and this may change urinary biochemical values. Some investigations require the use of preservatives or the samples may be collected over ice. Catheterization and random sampling can be used successfully in some studies.

3.1.9 Summary

Some estimates of the biological components of variation which occur in animal studies have been made (Lindena *et al.*, 1984; Carakostas and Banerjee, 1990). In general, the biological component is higher than the analytical component which is discussed in the next section. It is important to recognize the factors which contribute to the biological variation and to minimize these variables by careful study design. Failure to control these factors reduces the predictive value of the study and obfuscates 'real' toxic or pathological effects.

References

APOSTOLOU, A., SAIDT, L. & BROWN, W. R. (1976) Effect of overnight fasting of young rats on water consumption, body weight, blood sampling and blood composition. *Laboratory Animal Science*, **26**, 959–60.

BEAN-KNUDSEN, D. E. & WAGNER, J. E. (1987) Effect of shipping stress on clinicopathologic indicators in F344/N rats. *American Journal of Veterinary Research*, **48**, 306–8.

BERTANI, T., ZOJA, C., ABBATE, M., ROSSINI, M. & REMUZZI, G. (1989) Age-related nephropathy and proteinuria in rats with intact kidneys exposed to diets with different protein content. *Laboratory Investigations*, **60**, 196–204.

BOEMKE, W., PALM, U., KACZMARCZYK, G. & REINHARDT, H. W. (1990) Effect of high sodium and high water intake on 24-h-potassium balance in dogs. *Zeitschrift Versuchstierkunde*, **33**, 179–85.

CAISEY, J. D. & KING, D. J. (1980) Clinical chemistry values for some common laboratory animals. *Clinical Chemistry*, **26**, 1877–9.

CARAKOSTAS, M. C. & BANERJEE, A. K. (1990) Interpreting rodent clinical laboratory data in safety assessment studies: biological and analytical components of variation. *Fundamental and Applied Toxicology*, **15**, 744–53.

CHIN, B. H., TYLER, T. R. & KOZBELT, S. J. (1979) The interfering effects of hemolyzed blood on rat serum chemistry. *Toxicologic Pathology*, **7**, 19–21.

DANCLA, J. L. & NACHBAUR, J. (1981) The effect of diet on some clinical chemistry parameters in mice. *Journal of Clinical Chemistry and Clinical Biochemistry*, **119**, Abs.

DAVY, C. W., TRENNERY, P. N., EDMUNDS, J. G., ALTMAN, J. F. B. & EICHLER, D. A. (1987) Local myotoxicity of ketamine hydrochloride in the marmoset. *Laboratory Animals*, **21**, 60–7.

DROZDOWICZ, C. K., BOWMAN, T. A., WEBB, M. L. & LANG, C. M. (1990) Effect of in-house transport on murine plasma corticosterone concentration and blood lymphocyte populations. *American Journal of Veterinary Research*, **51**, 1841–6.

EVANS, G. O. (1985) Lactate dehydrogenase activity in platelets and plasma. *Clinical Chemistry*, **31**, 165–6.

(1987) Post-prandial changes in canine plasma creatinine. *Journal of Small Animal Practice*, **28**, 311–15.

(1994) Removal of blood from laboratory mammals and birds. *Laboratory Animals*, **28**, 178–9.

FALK, H. B., SCHROER, R. A., NOVAK, J. J. & HEFT, S. M. (1981) The effect of freezing on various serum chemistry parameters from common lab animals. *Clinical Chemistry*, **27**, 1039.

FOUTS, J. R. (1976) Overview of the field: environmental factors affecting chemical or drug effects in animals. *Federation Proceedings*, **35**, 1162–5.

FOWLER, J. S. L. (1982) Animal clinical chemistry and haematology for the toxicologist. *Archives of Toxicology*, **Suppl. 5**, 152–9.

FRIEDEL, K. & MATTENHEIMER, H. (1970) Release of metabolic enzymes from platelets during blood clotting of man, dog, rabbit and rat. *Clinica Chimica Acta*, **30**, 37–46.

GARATTINI, S. (1981) Toxic effects of chemicals: difficulties in extrapolating data from animals to man. *C.R.C. Critical Reviews in Toxicology*, **16**, 1–29.

GARNIER, F., BENOIT, E., VIRAT, M., OCHOA, R. & DELATOUR, P. (1990) Adrenal cortical response in clinically normal dogs before and after adaptation to a housing environment. *Laboratory Animals*, **24**, 40–3.

GARTNER, K., BUTTNER, D., DOHLER, K., FRIEDEL, R., LINDENA, J. & TRAUTSCHOLD, I. (1980) Stress response of rats to handling and experimental procedures. *Laboratory Animals*, **14**, 267–74.

HACKBARTH, H., BAUNACK, E. & WINN, M. (1981) Strain differences in kidney function of inbred rats: 1. Glomerular filtration rate and renal plasma flow. *Laboratory Animals*, **15**, 125–8.

HIRATA, M., NOMURA, G. & TANIMOTO, Y. (1979) Stability of serum components in monkey, dog and rat. *Experimental Animals*, **28**, 401–4.

JACOBS, R. M., LUMSDEN, J. H. & GRIFT, E. (1992) Effects of bilirubinemia, hemolysis and lipemia on clinical chemistry analytes in bovine, canine, equine and feline sera. *Canadian Veterinary Journal*, **33**, 605–7.

JENKINS, F. P. & ROBINSON, J. A. (1975) Serum biochemical changes in rats deprived of food and water for 24 h. *Proceedings of the Nutritional Society*, **34**, 37A.

JENSEN, A. L., WENCK, A. & KOCH, J. (1994) Comparison of results of haematological and clinical chemical analyses of blood samples obtained from the cephalic and external jugular veins in dogs. *Research in Veterinary Science*, **56**, 24–9.

KAST, A. & NISHIKAWA, J. (1981) The effect of fasting on oral acute toxicity of drugs in rats and mice. *Laboratory Animals*, **15**, 359–64.

KORSRUD, G. O. & TRICK, K. D. (1973) Activities of several enzymes in serum and heparinised plasma from rats. *Clinica Chimica Acta*, **48**, 311–15.

KUHN, G. & HARDEGG, W. (1988) Effects of indoor and outdoor maintenance of dogs upon food intake, body weight, and different blood parameters. *Zeitschrift für Versuchstierkunde*, **31**, 205–14.

LANDI, M. S. & KISSINGER, J. T. (1994) The effects of four types of restraint on serum alanine aminotransferase and aspartate aminotransferase in *Macaca fascicularis*. In *Welfare and Science, Proceedings of the Fifth FELASA Symposium*, pp. 37–40, London: Royal Society of Medicine Press.

LEVIN, S., SEMLER, D. & RUBEN, Z. (1993) Effects of two weeks of feed restriction on some common toxicologic parameters in Sprague–Dawley rats. *Toxicologic Pathology*, **21**, 1–14.

LINDENA, J., BUTTNER, D. & TRAUTSCHOLD, I. (1984) Biological, analytical and experimental components of variance in a long-term study of plasma constituents in rat. *Journal of Clinical Chemistry and Clinical Biochemistry*, **22**, 97–104.

LOEB, W. F. & QUIMBY, F. W. (1989) *The Clinical Chemistry of Laboratory Animals*. New York: Pergamon Press.

MAEJIMA, K. & NAGASE, S. (1991) Effect of starvation and refeeding on the circadian rhythms of hematological and clinico-biochemical values, and water intake of rats. *Experimental Animals*, **40**, 389–93.

MATSUZAWA, T., NOMURA, M. & UNNO, T. (1993) Clinical pathology reference ranges of laboratory animals. *Journal of Veterinary Medical Science*, **55**, 351–62.

MATSUZAWA, T. & SAKAZUME, M. (1994) Effect of fasting on haematology and clinical chemistry values in the rat and dog. *Comparative Haematology International*, **4**, 152–6.

McGOWAN, M. W., ARTISS, J. D. & ZAK, B. (1984) Description of analytical problems arising from elevated serum solids. *Analytical Biochemistry*, **142**, 239–51.

McGUILL, M. W. & ROWAN, A. N. (1989) Biological effects of blood loss: implications for sampling volumes and techniques. *ILAR News*, **31**, 5–18.

MEIJER, G. W., DE BRUIJNE, J. & BEYNEN, A. C. (1987) Dietary cholesterol-fat type combinations and carbohydrate and lipid metabolism in rats and mice. *International Journal of Vitamin and Nutritional Research*, **57**, 319–26.

MEYER, O. A., KRISTIANSEN, E. & WURTZEN, G. (1989) Effects of dietary protein and butylated hydroxytoluene on the kidneys of rats. *Laboratory Animals*, **23**, 175–9.

MITRUKA, B. M. & RAWNSLEY, H. M. (1977) *Clinical Biochemical and Hematological Reference Values in Normal Experimental Animals.* New York: Masson Publishing USA Inc.

MYERS, K. R. & MACKENZIE, K. M. (1985) A comparison of 24-hour urine output in male and female mice and rats. *Laboratory Animal Science*, **35**, 546–7.

NACHBAUR, J., CLARKE, M. R., PROVOST, J. P. & DANCLA, J. L. (1977) Variations of sodium, potassium and chloride plasma levels in the rat with age and sex. *Laboratory Animal Science*, **27**, 972–5.

NAKAMURA, M., ITOH, T., MIYATA, K., HIGASHIYAMA, N., TAKESUE, H. & NISHIYAMA, S. (1983) Difference in urinary N-acetyl-beta-D-glucosaminidase activity between male and female beagle dogs. *Renal Physiology (Basel)*, **6**, 130–3.

NEPTUN, D. A., SMITH, C. N. & IRONS, R. (1985) Effect of sampling site and collection method on variations in baseline clinical pathology parameters in Fischer 344 rats. I. Clinical chemistry. *Fundamental and Applied Toxicology*, **5**, 1180–5.

NIGI, H. & TORII, R. (1991) Periovulatory time courses of serum LH in the Japanese monkey (*Macaca fuscata*). *Experimental Animals*, **40**, 401–5.

OISHI, S., OISHI, H. & HIRAGA, K. (1979) The effect of food restriction for 4 weeks on common toxicity parameters in male rats. *Toxicology and Applied Pharmacology*, **47**, 15–22.

O'NEILL, S. L. & FELDMAN, B. F. (1989) Hemolysis as a factor in clinical chemistry and hematology of the dog. *Veterinary Clinical Pathology*, **18**, 58–68.

ORTH, D. N., PETERSON, M. E. & DRUCKER, W. D. (1988) Plasma immunoreactive proopiomelanocortin peptides and cortisol in normal dogs and dogs with Cushing's syndrome: diurnal rhythm and response to various stimuli. *Endocrinology*, **122**, 1250–62.

REIMANN, K. A., KNOWLEN, G. G. & TVEDTEN, H. W. (1989) Factitious hyperkalaemia in dogs with thrombocytosis. *Journal of Veterinary Internal Medicine,* **3**, 47–52.

REIMERS, T. J., LAMB, S. V., BARTLETT, S. A., MATAMOROS, R. A., COWAN, R. G. & ENGLE, J. S. (1991) Effects of hemolysis and storage on quantification of hormones in blood samples from dogs, cattle and horses. *American Journal of Veterinary Research*, **52**, 1075–9.

RILEY, V. (1981) Psychoneuroendocrine influence on immuno-competence and neoplasia. *Science*, **212**, 1100–2.

SCHEVING, L. E., SCHEVING, L. A., FEUERS, R. J., TSAI, T. H. & COPE, F. O. (1993). Chronobiology as it relates to Toxicology, Pharmacology and Chemotherapy. *Regulatory Toxicology and Pharmacology*, **17**, 209–18.

SCHWARTZ, E., TORNABEN, J. A. & BOXILL, G. C. (1973) The effects of food restriction on haematology, clinical chemistry and pathology in the albino rat. *Toxicology and Applied Pharmacology*, **25**, 515–24.

SONNTAG, O. (1986) Haemolysis as an interference factor in clinical chemistry. *Journal of Clinical Chemistry and Clinical Biochemistry*, **24**, 127–39.

STONARD, M. D., SAMUELS, D. M. & LOCK, E. A. (1984) The pathogenesis of nephrocalcinosis induced by different diets in female rats, and the effect on renal function. *Food and Chemical Toxicology*, **22**, 139–46.

STREET, A. E., CHESTERMAN, H., SMITH, G. K. A. & QUINTON, R. M. (1968) The effect of diet on blood urea levels in the beagle. *Journal of Pharmacy and Pharmacology*, **20**, 325–6.

SUBER, R. L. & KODELL, R. L. (1985) The effect of three phlebotomy techniques on hematological and clinical chemical evaluation in Sprague–Dawley rats. *Veterinary Clinical Pathology*, **14**, 23–30.

SWAIM, L. D., TAYLOR, H. W. & JERSEY, G. C. (1985) The effect of handling techniques on serum ALT activity in mice. *Journal of Applied Toxicology*, **5**, 160–2.

SZENCI, O., BRYDL, E. & BAJCSY, C. A. (1991) Effect of storage on measured ionized calcium and acid-base variables in equine, ovine, and canine venous blood. *Journal of the American Veterinary Medical Association*, **199**, 1167–9.

THOMPSON, M. B. & KUNZE, D. J. (1984) Polyethylene glycol-6000 as a clearing agent for lipemic serum samples from dogs and the effects on 13 serum assays. *American Journal of Veterinary Research*, **45**, 2154–7.

TORII, R. & NIGI, H. (1994) Hypothalamo-pituitary testicular function in male Japanese monkeys (*Macaca fuscata*) in non-mating season. *Experimental Animals*, **43**, 381–7.

TOTH, L. A. & JANUARY, B. (1990) Physiological stabilization of rabbits after shipping. *Laboratory Animal Science*, **40**, 384–7.

TULI, J. S., SMITH, J. A. & MORTON, D. B. (1995) Stress measurements in mice after transportation. *Laboratory Animals*, **29**, 132–8.

UCHIYAMA, T., TOKOI, K. & DEKI, T. (1985) Successive changes in the blood composition of the experimental normal beagle dogs accompanied with age. *Experimental Animals*, **34**, 367–77.

VADIEI, K., BERENS, K. L. & LUKE, D. R. (1990) Isolation-induced renal functional changes in rats from four breeders. *Laboratory Animal Science*, **40**, 56–9.

VOGEL, W. H. (1987) Stress – the neglected variable in experimental pharmacology and toxicology. *Trends in Pharmacological Sciences*, **8**, 35–8.

WHARY, M., PEPER, R., BORKOWSKI, G., LAWRENCE, W. & FERGUSON, F. (1993) The effects of group housing on the research use of the laboratory rabbit. *Laboratory Animals*, **27**, 330–41.

WOLFORD, S. T., SCHROER, R. A., GOHS, F. X., GALLO, P. P., BRODECK, M., FALK, H. B. & RUHREN, R. (1987) Age-related changes in serum chemistry and hematology values in normal Sprague–Dawley rats. *Fundamental and Applied Toxicology*, **8**, 80–8.

YOSHIDA, T., OHTOH, K., NARITA, H., OHKUBO, F., CHO, F. & YOSHIKAWA, Y. (1994) Feeding experiment on laboratory-bred Cynomolgus monkeys: II. Hematological and serum biochemical studies. *Experimental Animals*, **43**, 199–207.

Section 3.2 Analytical Variables

J. ROBINSON & G. O. EVANS

In this section, we discuss some of the analytical variables which require consideration when performing studies and interpreting data. The goal approved by the World Association of Societies of Pathology (1979) is that analytical imprecision should be equal to or less than one-half of the average within subject biological variation, i.e. analytical CV$<\frac{1}{2}$ biological CV (where CV is the coefficient of variation – see later). For commonly measured parameters in animal clinical chemistry, this goal is often achieved because the intrabiological variations (discussed in the previous section) are much greater than in human medicine.

The factors governing analytical procedures can be divided into the five 'M's: these factors are (1M) Manpower or analysts, (2M) Methods or analytical procedures, (3M) Machines or analytical equipment and reagents, (4M) Materials (samples, reference materials, standards or calibrants), and (5M) Manipulation of data for the calculation of results, reformatting of reported data and data transmission to host computers. Good Laboratory Practice demands attention to all of these M factors. Simple items such as reagent labelling, expiry dates and correct preparation often receive close attention during quality assurance inspections.

In most toxicology laboratories, there has been a continual increase in workload which has followed demand, with improvements in automated analysis both in test analysis productivity and the essential reduction of sample volumes required for analyses. Improvements of instrumental design now allow a greater number of tests to be performed with the samples obtained from laboratory animals. In addition to the development of analytical procedures, most of the current instrumentation incorporates software which enables data manipulation, calculation of results and 'real time monitoring' of the quality of data.

3.2.1 Methodology and Quality Control Procedures

The selection of appropriate instrumentation and methodology is an important step in controlling analytical performance. In general, most of the commercially

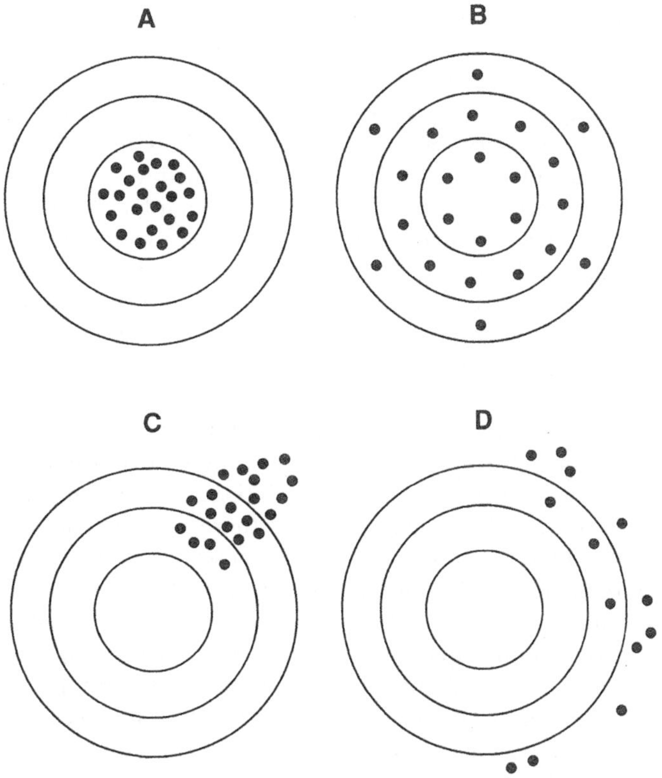

Figure 3.2.1 Target board figures illustrating accuracy and precision

available reagent kits and instrumentation are designed for use in human medicine, and any subsequent modifications for analysing samples from laboratory animals are left to the users. Knowledge of the biochemical principles and limitations for each of the methodologies is essential, if problems are to be avoided or at least recognized. Training and well-written Standard Operating Procedures help to minimize differences between analysts using the same analytical methods.

It is important to establish the accuracy and reproducibility of each method over the measuring range anticipated for each species. Linearity of a method should be established, or a series of standards selected for use with non-linear method calibration. Accuracy reflects the agreement between the test results and the true results, which in practice may be interpreted as showing agreement either with reference materials (including sera and urine) or primary standards which have been assayed by other laboratories. Very few primary standards are available for laboratory animals. These reference materials may be obtained from a commercial source or on a regular basis via an External Quality Assessment Scheme (EQAS).

Examples of methodological performance are indicated in target boards in Figure 3.2.1 where method A reflects an accurate and precise method, B an inaccurate and imprecise method, C a precise but inaccurate method, and method D an imprecise and inaccurate method.

Reproducibility, precision or perhaps more correctly imprecision of a method is usually expressed as the degree of agreement between replicate analysis on the same sample, and as a coefficient of variation (CV) where the standard deviation, which is a measure of the dispersion of a group of values around the mean, is expressed as a percentage of the mean. Values for the CV can be determined for within-batch or analytical run and between-batch or day.

For common analytes, the within-batch and between-batch CV values are often less than 5% with current analytical methods and instrumentation. The CV values are slightly higher, indicating greater imprecision for immunoassays, e.g. hormones, and where methods are scaled down to deal with smaller sample volumes; manual methods are generally less precise than automated methods. Thus, analytical variations are relatively low in comparison to other variables within a toxicology study.

Amongst the tests required by regulatory authorities discussed in Chapter 2 were gamma glutamyl transpeptidase (GGT) and total bilirubin. In most rats, plasma GGT activity is very low (usually <3 IU/L) and plasma bilirubin values are in the range 0–4 μmol/l. In many studies there are no significant changes of these two plasma analytes, although both of these parameters may change with hepatotoxicity (Chapter 6). As both of these measurements are often made where the method is imprecise and at the limits of sensitivity for the assays, small changes of absorbance can result in apparent effects on the plasma levels: plasma GGT may change from 1 to 3 IU/L and similar changes may occur for plasma bilibrubin. These small changes can result in significant statistical differences when comparing treatment groups or pre- and post-treatment values, but consideration of the analytical component can lead to the opinion that there are no differences due to treatment.

Various quality control procedures are used by laboratories to ensure reliability, and to reduce or eliminate variability caused by instrumentation and/or reagents. Analytical errors can occur randomly or be systematic. Whilst random errors such as incorrectly identified samples or incorrect loading of samples onto the analyser cannot be easily found, systematic errors can be recognized by quality control procedures. Inevitably in a large data set, there will be some values which do not agree with the distribution of the majority of values – so-called outliers. Here, a problem is to identify if these values are 'true' biological variations or a result of random analytical errors.

Each analytical batch should contain quality control materials (serum, urine or a suitable aqueous control). These materials can be used to monitor both precision and accuracy and preferably with at least two different concentration levels (low and high). Quality control data obtained on a daily, weekly or monthly basis can be examined with a variety of techniques including the Levey-Jennings plots, cusum plots and Westgard rules (Westgard and Klee, 1994). These techniques are used to detect relative movements of precision and accuracy for a given method, and thus allow corrective action to be taken.

Most laboratories participate in an External Quality Assessment Scheme (EQAS) where samples are distributed on a regular basis to participating laboratories, and results are compared to overall or group mean values as an indicator of performance standards.

3.2.2 Other Examples of Analytical Variables

Sample Evaporation

Even with the sophistication of today's analysers, samples may evaporate during the analytical procedures and this will affect results (Burtis *et al.*, 1975; Salzmann and Male, 1993). This evaporation problem is particularly important in the analysis of small volume samples taken from laboratory animals.

Lipids

Methods for plasma cholesterol and triglycerides use various enzymes for lipid hydrolysis in the initial stages of the reactions, and several examples of differences and underestimation of either cholesterol or triglycerides have been reported (Chapter 13).

Proteins

Variations of dye-binding with albumin from different species have been described. The commonest dye-binding method for plasma albumin uses bromocresol green, and it is essential to make absorbance measurements within 60 s if overestimation of albumin resulting from other reactions with alpha-1-globulin fractions is to be avoided. Methods using bromocresol purple suggested as an alternative for bromocresol green underestimate plasma albumin levels unless species-specific standards are used, and very few of the animal proteins are available and sufficiently pure to be used as primary calibrants. Alterations of electrophoretic apparatus, support media and buffers will alter the apparent number of protein fractions obtained on separation (Chapter 14).

Creatinine

Methods based on the reaction of creatinine with alkaline picrate are commonly used, although it is recognized that these methods are subject to many interferences and the values for creatinine include a significant proportion due to non-creatinine chromogens. For example, serum creatinine values in mice obtained with an alkaline picrate method were fivefold the values determined by a high-performance liquid chromatographic (HPLC) method (Meyer *et al.*, 1985). Using an enzymatic method, Evans (1985) demonstrated that the alkaline picrate method overestimated creatinine in several species.

Enzymes

Dooley (1979) reported the optimum substrate concentrations required for the measurement of plasma aspartate aminotransferase (AST/GOT) in rats, dogs and monkeys were markedly different from the concentrations used for human plasma assays. Although reagents for enzyme measurement are known not to be optimal for the different species, there is a lack of data on the optimal conditions – buffer, substrate concentration etc. – required for the different species and there are no nationally or internationally agreed recommendations for laboratory animal samples.

Hormones

See Chapter 8.

3.2.3 Changing Analytical Methods

Although assay systems can be modified for use with a particular species, this is not always easily achieved. The problem for many manufacturers is that the commercial market is too small to justify the expenditure required to develop reagents which are species-specific. Problems associated with lipid measurements have shown the need for analysts to be vigilant when new reagent formulations are introduced.

Recognizing the limitations of certain methods in terms of accuracy is important, and it must be remembered that the analytical limitations will apply to pre- and post-treatment samples, and to control groups as well as treatment groups. In Section 3.3, we consider the effects of the presence of test compound (and/or its metabolites) on the analytical methods.

Several methodological changes which occur during a study may introduce a bias to the data; for example, in longer-term 2-year studies where new instrumentation or reagent formulations are introduced into the laboratory. In some situations it may be possible to continue with the original methods until completion of the study. Problems also may occur where an instrument fails to operate correctly, and a 'back-up' method using slightly different instrumentation or reagents is employed. Variations in analytical methodologies and units can often lead to problems when data are compared for the same test compound in different toxicological studies obtained from several laboratories. In all of these situations, it is useful to know the analytical bias which may have been introduced by such methodological changes, and to take these factors into account when interpreting data.

References

BURTIS, C. A., BEGOVICH, J. M. & WATSON, J. S. (1975) Factors influencing evaporation from sample cups and assessment of their effect on analytical error. *Clinical Chemistry*, **21**, 1907–17.

DOOLEY, J. F. (1979) The role of clinical chemistry in chemical and drug safety evaluation by use of laboratory animals. *Clinical Chemistry*, **25**, 345–7.

EVANS, G. O. (1985) The use of an enzymatic kit to measure plasma creatinine in the mouse and three other species. *Comparative Biochemistry and Physiology*, **85B**, 193–5.

MEYER, M. H., MEYER, R. A., GRAY, R. W. & IRWIN, R. L. (1985) Picric acid methods greatly overestimate serum creatinine in mice: more accurate results with high-performance liquid chromatography. *Analytical Biochemistry*, **144**, 285–90.

SALZMANN, M. B. & MALE, I. A. (1993) Keep the lid on it: artefactual hypernatraemia in samples from paediatric patients. *Annals of Clinical Biochemistry*, **30**, 211–12.

WESTGARD, J. O. & KLEE, G. G. (1994) Quality management. In Burtis, C. A. & Ashwood, E. R. (eds) *Tietz Textbook of Clinical Chemistry*, 2nd Edn, pp. 548–93. Philadelphia: W. B. Saunders.

WORLD ASSOCIATION OF SOCIETIES OF PATHOLOGY (1979) Proceedings of the Subcommittee on Analytical Goals in Clinical Chemistry. Analytical goals in clinical chemistry: their relationship to medical care. *American Journal of Clinical Pathology*, **71**, 624–30.

Section 3.3 Analytical Interference

G. O. EVANS

When examining data from toxicological studies, differences are sometimes found for clinical chemistry data which cannot be explained simply on the grounds of observed toxicity, histopathological findings, clinical observations, statistical analyses, or preanalytical variables. In these cases it may be worth considering whether the differences are due to the effect(s) of the test compound and/or its metabolites on the assay(s) in question, and this general consideration of metabolites applies throughout the following section where only the test compound may be mentioned.

Xenobiotics may interfere with analytical methods in several different ways and these may have a negative or positive effect on the result: these effects may be classified as biological or analytical. Biological effects of drugs include the desirable primary therapeutic effect, e.g. the effects of thiazide diuretics, potassium supplements to treat hypokalaemia, and allopurinol on urate metabolism. Secondary biological effects or side effects may be undesirable, e.g. changes of plasma bilirubin with phenobarbitone administration, elevated plasma amylase following the contraction of the sphincter of Oddi by morphine, and coagulation proteins with oral contraceptives.

An interfering substance in analytical procedures causes a systematic error in the analytical results at a given concentration. A test compound causing the interference is sometimes called an interferent. The extent of the interference will obviously depend on the level of interferent in the test samples, and this effect may be exaggerated in toxicology studies where the levels may be several multiples of the therapeutic dosages. This form of interference occurs when the test compound is measured directly as if it were the analyte itself, or through its influence as an accelerator or inhibitor of the chemical reactions used to measure the analyte. These methodological interferences are more liable to occur in analytical procedures which use non-specific reactions.

Physical interferences often arise when the test compound or metabolite has the same colour as the end-point colour of the reaction used to measure the analyte, or when the sample is coloured by haemolysis which particularly affects colorimetric methods using absorbance reading at 540 nm. Other xenobiotics may interfere because of their fluorescent properties.

3.3.1 Examples of Interference

3.3.1.1 Preanalytical

In the previous sections of this chapter some of the possible effects due to preanalytical factors were considered. Endogenous causes of interference include haemolysis, lipaemia, bilirubinaemia and paraproteinaemia (Kroll and Elin, 1994). Another problem may be introduced by using one of the several polymeric materials which are used as serum separator devices. These polymeric materials have a density between that of serum and the blood cells and thus form a layer on top of the packed cells during centrifugation. When using these separators, small differences have been reported, e.g. lower potassium values for several species (Caisey and King, 1980), but Sunaga *et al.* (1992) reported no marked differences for several analytes using three different separators with several species. There is evidence that some of these serum separators may have a potential to interact with hydrophobic drugs in human serum (Landt *et al.*, 1993).

Some other effects include inappropriate anticoagulant, e.g. potassium interference associated with sequestrenated blood, iron chelators such as desferroxamine with iron, and effects due to intramuscular injections (see Chapter 10 – Cardiotoxicity and Myotoxicity).

3.3.1.2 Plasma Protein Binding

Xenobiotics can bind to the various plasma protein fractions and the degrees to which this binding occurs can have a marked effect on the pharmacological action of a drug (Goldstein *et al.*, 1974; Lindup and L'E Orme, 1981). There are many instances where xenobiotics bind to plasma proteins such as albumin but there are surprisingly few instances where this has been reported to interfere with the determination of plasma albumin. However, the binding of a test compound to proteins can have marked effects on other laboratory tests. This is exemplified by the effect on prothrombin time following the displacement of coagulation proteins by phenylbutazone, and the action of drugs on thyroid hormones (Chapter 8).

3.3.1.3 Interference in Cascade Methods

There are several methods where initial reactions with the test analyte are followed by the measurement of hydrogen peroxide via peroxidases linked to a colorimetric end point. Examples of this type of interference are seen in peroxidase methods with 4-aminophenazone and phenol which yield quinoneimine as chromogen (Trinder, 1969; Fossati *et al.*, 1980), and several methods for diverse plasma constituents including glucose, urate, cholesterol, and triglycerides use this or similar end-point reactions. Thus we have methods where relatively specific enzyme-mediated reactions are used in the first stage, but the subsequent stages are subject to interference from a variety of compounds.

3.3.2 Data Banks

Several collections of reported interferences of laboratory tests by drugs have now been followed by more systematic data banks (Caraway, 1962; Wirth and Thompson, 1965; Elking and Kabat, 1968; Lubran, 1969; van Peenan and Files, 1969; Christian, 1970; Sher, 1982). Young *et al.* (1975) reported nearly 20 000 interference effects. Other more selective databases have subsequently been established (Salway, 1990; Young, 1990; Tryding *et al.*, 1992). In most of these data banks, the analytical interferences with clinical chemistry tests account for the minority of reported interferences due to drugs. Whilst these databases are primarily designed to help the physician, they can be used as an aid in identifying possible mechanisms for interferences with test compounds.

3.3.3 Testing for Analytical Interference

Various guidelines have been proposed for testing for drug interference in clinical chemistry measurements and they suggest the necessity for performing *in vitro* studies (Powers *et al.*, 1986; Kallner and Tryding, 1989; Kroll and Elin, 1994), but there are several limitations to performing such studies. *In vitro* studies may not always be relevant to clinical practice, e.g. high levels of ascorbic acid interfere in several measurements but are rarely seen *in vivo*. More importantly xenobiotics are often transformed into a wide range of metabolites, and consideration then has to be given to the relative *in vivo* abundance of the metabolite and its availability for testing. Concurrent administration of other drugs can complicate interpretation of effects due to interference.

Some of the problems of testing for interferences are shown by studies with several cephalosporins which interfere with creatinine measurements. The measurement of creatinine is important in monitoring renal function following administration of cephalosporin antibiotics, and it is commonly measured with alkaline picrate reagents. Although the methodological principle is common, cephalosporins can interfere either positively or negatively with serum creatinine measurements depending on the instrumentation and reaction conditions (Kirby *et al.*, 1982; Okudo *et al.*, 1984; Kroll *et al.*, 1985; Letellier and Desjarlais, 1985b; Appel *et al.*, 1991). Durham *et al.* (1979) showed that by altering the timing of the blood sample in relation to dosing with cefoxitin, the interference could be reduced as the plasma drug levels fell.

Baer *et al.* (1983) performed 12 000 measurements in a study of the effect of cefotaxime and one of its major metabolites on 24 common clinical chemistry tests using one analyser. Other similar studies have been carried out with ascorbic acid, analgesics and anti-rheumatic drugs (Siest *et al.*, 1978; Jelic-Ivanovic *et al.*, 1985a, 1985b). From a study of 20 drugs with 7 different instruments, Letellier and Desjarlais (1985a) concluded that interferences change in magnitude and direction depending on the instrumentation and reaction conditions. Given the variety of analytical methods and instrumentation available, it is not possible to test every combination, and it is important to recognize that these factors change with time (Evans, 1985). Some proposals for defining clinically significant interference as compared to significant effects on the analyte have been made (Fuentes-Arderiu and Fraser, 1991; Castano-Vidriales, 1994).

The proposals for *in vitro* interference testing are not always practicable, and they should not be a first-line consideration when performing toxicological studies where the priority is to identify target organ toxicity. *In vitro* studies can be useful when testing series of compounds which are similar in chemical structure and/or therapeutic target and where there has been some association with marked interference effects. It is sometimes useful to perform the analysis using a method with a different principle where interference is suspected.

3.3.4 Summary

Interference effects may be biological, analytical, physical or chemical in nature and vary for the same analyte when different analytical methods and/or instrumentation are employed. The concentrations of the test compound and/or metabolite(s) in blood and urine must be considered in deciding if the effects are due to interference(s) with the assay, and additional problems occur when more than one compound is being tested. Following *in vivo* studies, testing the test compound and/or a metabolite by *in vitro* techniques can be useful. Early recognition of potential interference in plasma or urine chemistry measurements during the preclinical or development phase of a test compound, however, can assist and prevent misleading interpretations made during later stages of development results.

References

Appel, W., Hubbuch, A. & Koller, P. U. (1991) *In-vitro*-Profung auf interfernzen durch Pharmaka bei quantitativen Urinanalysen: Ergebnisse eines Expertgesprachs und erste Erfahrungen mit der neuen Prufliste. *Laboratorium Medicine*, **15**, 399–403.

Baer, D. M., Jones, R. N., Mullooly, J. P. & Horner, W. (1983) Protocol for the study of drug interference in laboratory tests: cefotaxime interference in 24 clinical tests. *Clinical Chemistry*, **29**, 1736–40.

Caisey, J. D. & King, D. J. (1980) Clinical chemical values for some common laboratory animals. *Clinical Chemistry*, **26**, 1877–9.

Caraway, W. T. (1962) Chemical and diagnostic specificity of laboratory tests. *American Journal of Clinical Pathology*, **37**, 1453–8.

Castano-Vidriales, J. L. (1994) Interferences in clinical chemistry. *Journal of the International Federation of Clinical Chemistry*, **6**, 10–14.

Christian, D. G. (1970) Drug interference with laboratory blood chemistry determinations. *American Journal of Clinical Pathology*, **54**, 118–42.

Durham, S. R., Bignell, A. H. C. & Wise, R. (1979) Interference of cefoxitin in the creatinine estimation and its clinical relevance. *Journal of Clinical Pathology*, **32**, 1148–51.

Elking, P. & Kabat, H. F. (1968) Drug induced modifications of laboratory test values. *American Journal of Hospital Pharmacy*, **25**, 485–519.

Evans, G. O. (1985) Changes of methodology and their potential effects on data banks for drug effects on clinical laboratory tests. *Annals of Clinical Biochemistry*, **22**, 397–401.

Fossati, P., Principe, L. & Berti, G. (1980) Use of 3.5 dichloro-2-hydroxybenzene sulphonic acid-4-aminophenazone chromogenic system in direct assay of uric acid in serum and urine. *Clinical Chemistry*, **26**, 227–31.

FUENTES-ARDERIU, X. & FRASER, C. G. (1991) Analytical goals for interference. *Annals of Clinical Biochemistry*, **28**, 393–5.

GOLDSTEIN, A., ARONOW, L. & KALMAN, S. M. (1974) *Principles of Drug Action*, pp. 158–64. New York: J. Wiley.

JELIC-IVANOVIC, Z., SPASIC, S., MAJKIC-SINGH, N. & TODOROVIC, P. (1985a) Effects of some anti-inflammatory drugs on 12 blood constituents: protocol for the study of *in vivo* effects of drugs. *Clinical Chemistry*, **31**, 1141–3.

JELIC-IVANOVIC, Z., MAJKIC-SINGH, N., SPASIC, S., TODOROVIC, P. & ZIVANOV-STAKIC, D. (1985b) Interference by analgesic and antirheumatic drugs in 25 common laboratory assays. *Journal of Clinical Chemistry and Clinical Biochemistry*, **23**, 287–92.

KALLNER, A. & TRYDING, N. (1989) IFCC Guidelines to the evaluation of drug effects in Clinical Chemistry. *Scandinavian Journal of Clinical and Laboratory Investigation*, **49**, 1–29.

KIRBY, M. G., GAL, P., BAIRD, H. W. & ROBERTS, B. (1982) Cefoxitin interference with serum creatinine measurement varies with the assay system. *Clinical Chemistry*, **28**, 1981.

KROLL, M. H. & ELIN, R. J. (1994) Interference with clinical laboratory analyses. *Clinical Chemistry*, **40**, 1996–2005.

KROLL, M. H., NEALON, L., VOGEL, M. A. & ELIN, R. J. (1985) How certain drugs interfere negatively with the Jaffe reaction for creatinine. *Clinical Chemistry*, **31**, 306–8.

LANDT, M., SMITH, C. H. & HORTIN, G. L. (1993) Evaluation of evacuated blood-collection tubes: effects of three types of polymeric separators on therapeutic drug monitoring specimens. *Clinical Chemistry*, **39**, 1712–17.

LETELLIER, G. & DESJARLAIS, F. (1985a) Analytical interference of drugs in clinical chemistry: 1. Study of twenty drugs on seven different instruments. *Clinical Biochemistry*, **18**, 345–51.

(1985b) Analytical interference of drugs in clinical chemistry: II. The interference of three cephalosporins with the determination of serum creatinine concentration by the Jaffe reaction. *Clinical Biochemistry*, **18**, 352–6.

LINDUP, W. E. & L'E ORME, M. C. L. (1981) Plasma protein binding of drugs. *British Medical Journal*, **282**, 212–14.

LUBRAN, M. (1969) The effects of drugs on laboratory values. *Medical Clinics of North America*, **53**, 211–22.

OKUDO, T., ITO, J. & NISHIDA, M. (1984) 'Interferovalue' indicates the interference of substances with creatinine determination. *Clinical Chemistry*, **30**, 1888–9.

POWERS, D. M., BOYD, J. C., GLICK, M. R., KOTSCHI, M. L., LETELLIER, G., MILLER, W. G., NEALON, D. A. & HARTMANN, A. E. (1986) Interference testing in clinical chemistry: proposed guidelines. *NCCLS Document EP7-P*. Villanova, PA: National Committee for Clinical Laboratory Standards.

SALWAY, J. G. (1990) *Drug Test Interactions Handbook*. London: Chapman & Hall.

SHER, P. P. (1982) Drug interferences with clinical laboratory tests. *Drugs*, **24**, 24–63.

SIEST, G., APPEL, W., BLIJENBERG, G. B., CAPOLAGHI, B., GALTEAU, M. M., HEUGSHEM, C., HJELM, M., LAUER, K. L., LE PERRON, B., LOPPINET, V., LOVE, C., ROYER, R. J., TOGNONI, C. & WILDING, P. (1978) Drug interference in clinical chemistry: studies on ascorbic acid. *Journal of Clinical Chemistry and Clinical Biochemistry*, **16**, 103–10.

SUNAGA, T., HIRATA, M., ICHINOHE, K., SAITO, E., SUZUKI, S. & TANIMOTO, Y. (1992) The influence of serum separators on biochemical values in experimental animals. *Experimental Animals*, **41**, 533–6.

TRINDER, P. (1969) Determination of glucose in blood using glucose oxidase with an alternative oxygen acceptor. *Annals of Clinical Biochemistry*, **24**, 24–7.

TRYDING, N., HANSSON, P., TUFVESSON, C., SJOLIN, T. & SONNTAG, O. (1992) *Drug Effects in Clinical Chemistry*, 6th Edn. Stockholm: Apoteksbolaget Lakemedelsverket and Svensk Forening fur Klinisk Kemi.

VAN PEENAN, H. J. & FILES, J. (1969) The effect of medication on laboratory test results. *American Journal of Clinical Pathology*, **52**, 666–70.

WIRTH, W. A. & THOMPSON, R. L. (1965) The effect of various conditions and substances on the results of laboratory procedures. *American Journal of Clinical Pathology*, **43**, 579–90.

YOUNG, D. S. (1990) *Effects of Drugs on Clinical Laboratory Tests*. Washington: AACC Press.

YOUNG, D. S., PESTANER, L. C. & GIBBERMAN, V. (1975) Effects of drugs on clinical laboratory tests. *Clinical Chemistry*, **21**, 1D-432D.

4

Statistical Approaches

A. DICKENS & J. ROBINSON

There are three kinds of lies – Lies, Damn Lies and Statistics.

(Benjamin Disraeli)

4.1 Introduction

The problem with statistics is that they are often misinterpreted: for example, if 30% of all accidents are caused by drunken drivers, it does not imply that 70% are caused by sober ones, who are therefore the greater danger: so you have to think about what the figures really mean. The death rate in the American Navy during the Spanish–American War was 0.9%. For civilians in New York City during the same period it was 1.6%. This suggests that it was safer to be in the Navy than out of it, but the groups are not really comparable. The Navy is made up of mostly young healthy men, whereas the civilian population includes infants, the old, and the ill, all of whom have a higher death rate wherever you are.

It is easy to see with the above simple examples how the use and interpretation of statistics are misunderstood, without delving into more complex research problems. Such views have in the past helped to build up an aversion to statistics which is even now considerable, despite the rapid advances and the contributions which statistical methods have been able to make. The first part of this chapter is concerned mainly with some of the basic tools for exploratory analysis and presentation, a part of the subject usually called **descriptive statistics**. The data from experiments should always be examined before any formal analysis is performed. Such examinations should occur at the initiation of an analysis (since the raw data will often be too extensive and too complex to convey an immediate impression) and during the analysis to improve the decisions on what form the analysis should take.

It is essential that every experiment yields as much information as possible, and that each experiment's results have the greatest possible chance of answering the questions it was conducted to address. The statistical aspects of such efforts, so far as they are aimed at structuring experiments to maximize the possibilities of success, are called **experimental design**. The second part of this chapter gives a brief discussion of a few of the general principles involved in the design of experiments.

In the final part of the chapter, different statistical techniques that might generally be applied to different types of toxicological data are summarized.

4.2 Descriptive Statistics

4.2.1 Diagrams

It is said that a picture is worth a thousand words. Certainly in newspapers, messages which words cannot convey can be very forcefully presented by pictures. In this respect, the use of statistical diagrams is an important area (particularly with the availability of computer packages) and there has been something of a resurgence of interest in diagrams.

Trends and contrasts are often more easily understood by casual inspection of a quality diagram than by scrutiny of a table of the corresponding numerical data. Diagrams must, however, be simple. If too much information is presented, it becomes difficult to interpret and people are unlikely even to make the effort to understand. Details are usually lost when data are displayed in diagrammatic form, thus reference must be made to the relevant numerical quantities for any critical analysis of the data.

Statistical diagrams serve two main purposes. The first is the use of diagrams to gain insight into the structure of the data, and to check assumptions which might be made in the analysis. This informal use of diagrams will often reveal new aspects of the data, or suggest further hypotheses which may be investigated. The second main use is in the presentation of statistical results in reports (Figure 4.1), where it may be felt that a simple, evocative display will be appreciated. The powerful impact of diagrams makes them a potential means of misrepresentation by the uneducated and the unscrupulous. For example, little attention should be paid to a diagram unless the variables and the scales are shown and clearly defined.

Figure 4.2a shows the trend of plasma LDH over a dose range. The figure shows a slight rise in LDH. The reader who glances quickly at the diagram may receive the impression that LDH increases dramatically. This illusion arises because the diagram is not complete. The y-axis does not start at zero and we are seeing only the top half of the picture. The whole of a picture must be seen to enable us to understand its full value. A very different presentation of the data is shown in Figure 4.2b, in which it appears that the differences are almost negligible. Nevertheless, there was a difference to some degree: this difference is grossly exaggerated in Figure 4.2a but is almost 'lost' in Figure 4.2b.

This illustrates at once the absolute necessity for not taking things for granted and for not reading what you expect to see into a diagram, without really making certain of the facts. Memory of past results similar to that apparently produced in a chart is sometime apt to make the reader susceptible to suggestion and to see something which is not there.

Two types of diagrams which display the overall pattern in a set of observations, the distribution, are shown in Figures 4.3 and 4.4. For a more detailed description of statistical diagrams reference should be made to Huff (1954) and Tukey (1977).

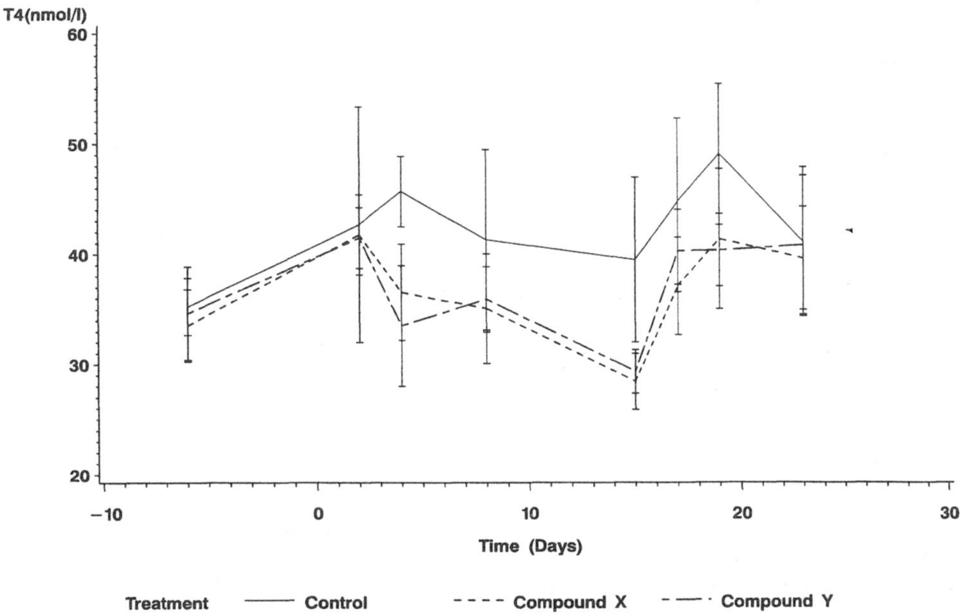

Figure 4.1 Plasma thyroxine (T4) in rodents after 15 days' treatment with either compound X or Y followed by 8 days' recovery

A **histogram** breaks the range of values of a variable into intervals and a rectangle is drawn above each class-interval (Figure 4.3) such that the area of each rectangle is proportional to the frequency of the observations falling in the corresponding interval. If intervals are of the same width then the height of the rectangles is proportional to the frequency, but it is not always practicable to make all the intervals the same width. For example, suppose we decide to pool the groups 121–40, 141–60 and 161–80. The total frequency in these groups is 12, but it would clearly be misleading to represent this frequency by a rectangle on a base extending from 121 to 180 and with a height of 12. The correct procedure would be to make the height of the rectangle 4, the average frequency in the three groups (as indicated by the dotted line in Figure 4.3).

Before drawing a histogram you must decide how many classes to use. This is a matter of judgement. Too few classes will group most observations together, while too many will place only a few observations in each class. Narrow classes preserve more detail but the heights of the bars often vary irregularly. Wide classes often give a more regular picture of the overall shape, but they lose detail by grouping a wider interval of values into each class. Neither extreme will show clearly the shape of the distribution. Fortunately, a broad range of choices will usually give a similar impression of the distribution.

The **stem and leaf plot** is a quick way to picture the shape of a distribution while at the same time including the actual numerical values in the diagram.

To make a stem and leaf plot:

1 Separate each observation into a stem and a leaf; in general, stems may have as many digits as needed, but each leaf should contain a single digit. For example, if you have a value of 148, the stem would be 14 and the leaf 8.

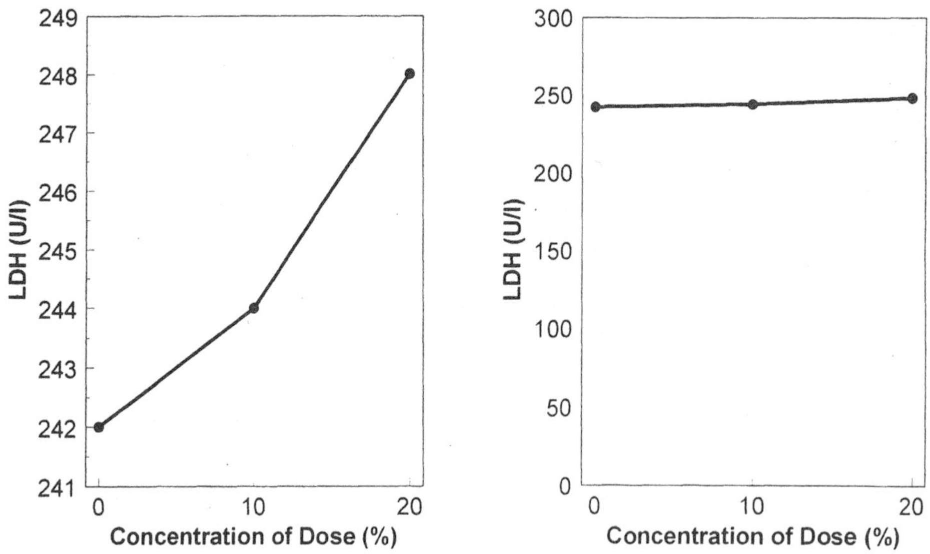

Figure 4.2a and b Effect of an inappropriate vertical axis scale

2 List the stems in vertically decreasing order from top to bottom, draw a vertical line to the right of the stems, and add the leaves to the right of the line.
3 Arrange the leaves in increasing order from left to right.

Thus, if we use the same data as for the histogram (Figure 4.3) the resulting stem and leaf plot would be as shown in Figure 4.4.

Once a distribution has been displayed by a histogram or stem and leaf plot several important features can be seen:

1 Locate the **centre of the distribution**, either by eye or by counting in from either end of the stem and leaf plot until half the observations are counted. This simple measure of the centre of a distribution is called the median (see next section for a formal description). In this example the median is 105.
2 Examine the overall **shape of the distribution**. Is it unimodal (one peak) or multimodal (many peaks)? Is it approximately symmetric or skewed in one direction? A distribution is symmetric if the portions above and below the median are roughly mirror images of each other. The general shape of symmetric and skewed distributions are shown in Figure 4.5.

   Thus, in the example it is clear that we have a positively skewed distribution.
3 Finally, look for marked deviations from the overall shape. There may be gaps in the distribution, or there may be **outliers**, individual observations that fall well outside the overall pattern of the data. If outliers are known to be atypical and not a fair representation of the treatment it is then permissible to reject the observations and to regard them as 'missing' in any analysis. Sometimes an observation seems very different from the others on the same treatment, though no reason can be given why this should be so. Such an observation

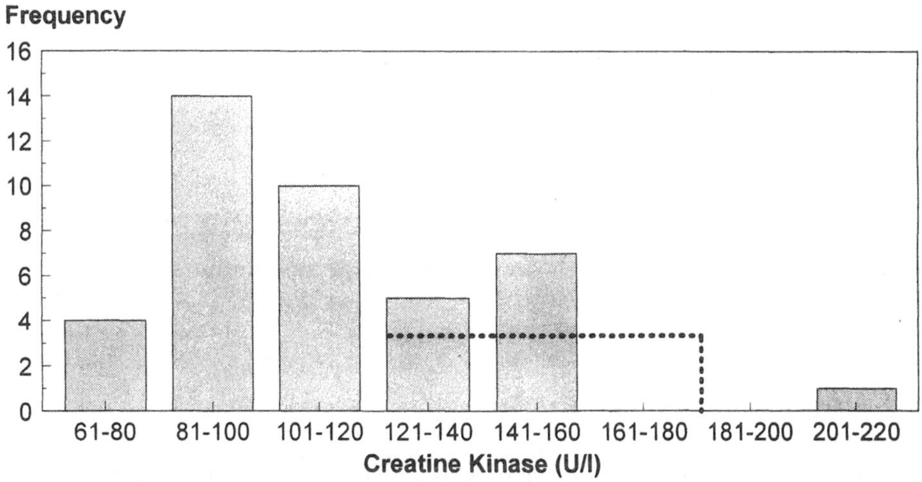

Figure 4.3 Histogram of plasma creatine kinase

20	9
19	
18	
17	
16	
15	2 8
14	1 4 6 7 9
13	4 4
12	1 1 3
11	3 4 6 7 9
10	1 3 5 7 8
9	2 5 5 6 6 8 8
8	1 2 5 6 7 8 8
7	0 3 8
6	9

Figure 4.4 Stem and leaf plot of plasma creatine kinase

should not be rejected unless the chance of it occurring is very small, and the fact that it has been rejected should be mentioned in any report of the experiment. Remember, once you start hand-selecting your results your sample becomes biased. The ultimate folly of rejecting an extreme observation was demonstrated when shortly after 7 o'clock on the morning of December 7, 1941, the officer in charge of a Hawaiian radar station ignored data solely because it seemed so incredible: this of course referred to the bombing raids on Pearl Harbor.

Making a graphical display is the first step towards understanding data; performing calculations which give the location of the distribution and the degree of variation of the observations is the second step.

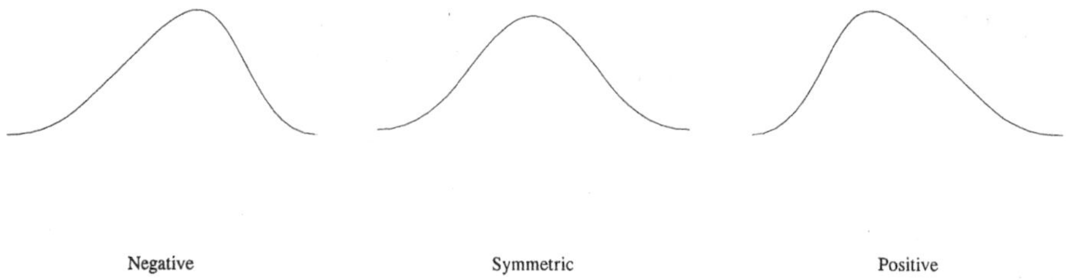

Figure 4.5 Illustration of negatively skewed, symmetric, and positively skewed distributions

4.2.2 Measure of Location

It is often important to give, in a single figure, some indication of the general level of a series of measurements. Such a figure may be called a measure of location. There are two commonly used measures of location, the mean and the median.

The **mean** of a population is denoted by μ (the Greek letter 'mu'), and the mean of a sample, containing n individuals, whose values are referred to as $x_1, x_2, \ldots, x_n$ by $\bar{x}$ (spoken 'x bar'). The mean is the sum of the observations divided by the number of observations. Thus,

$$\bar{x} = \frac{x_1 + x_2 + \ldots + x_n}{n} = \frac{\Sigma x_i}{n}$$

where the summation sign Σ means 'add them all up'. Thus if the experimenter measures the following LDH (U/1) values:

163, 140, 195, 133, 172, 141, 218, 652, 265, 183

the mean is:

$$\bar{x} = \frac{163 + 140 + \ldots + 183}{10} = \frac{2262}{10} = 226.2$$

Another useful measure of location is the **median**. To compute the median of a distribution:

1 Arrange all the observations in order of size, from the smallest to largest.
2 If the number of observations, n, is odd, the median is the centre, ½(n + l)th, observation in the ordered list.
3 If the number of observations, n, is even, the median is the mean of the two centre, ½nth and (½n + l)th, observations in the ordered list.

To find the median LDH value, we arrange the data in increasing order:

133, 140, 141, 163, 172, 183, 195, 218, 265, 652

Since the number of observations, n = 10, is even, the median is the mean of the two centre observations.

$$Median = \frac{172 + 183}{2} = 177.5$$

The median takes no account of the precise magnitude of most of the observations, and is therefore usually less efficient than the mean, because it wastes information. However, the mean can be misleading. If we consider the previous example, owing to the skewed nature of the distribution the mean of 226.2 U/L is not really typical of the series as a whole, and the median of 177.5 U/L might be a more useful index. This illustrates an important weakness of the mean as a measure of centre, in that it is sensitive to the influence of a few extreme values. Similarly, a mean should not be used as a measure of centre for skewed distributions as its value is pulled towards the long tail of the distribution. The median is therefore more stable than the mean in the sense that it is less likely to fluctuate from one series of readings to another.

4.2.3 Measures of Variation

Even the briefest summary of a distribution requires an indication of how spread out, or variable, the data are in addition to a measure of their centre. Are the values close to the centre or are some scattered wildly in each direction? This question is important for purely descriptive reasons and also since the measurement of variation plays a central part in the methods of statistical inference. We therefore require what is variously termed a measure of variation, scatter, spread or dispersion.

One way to measure this variability is simply to look at the highest and lowest of the observations and calculate the difference between them, the **range**. Note that the range is a definite quantity, measured in the same units as the original observations; thus we may say that the readings range from the lowest (133 U/L) to the highest (652 U/L) in the LDH series and the range is 519 U/L. The range is determined by two of the original observations, the lowest and the highest in the series, and ignores the pattern of distribution of the observations in between; thus as large differences in these extreme values are liable to occur in studies, the range can be unreliable and uninformative.

If the number of observations is not too small a modification may be introduced which avoids the use of these absolute extreme values. If the values are arranged in increasing order, two values may be ascertained which cut off a small fraction of the observations at each end, just as the median breaks the distribution into equal parts. The **first quartile** or **lower quartile** is the median of the lower 50% of observations, and the **third quartile** or **upper quartile** is the median of the upper 50% of observations. The quartiles together with the median give some indication of centre, spread and shape of a distribution. The distance between the quartiles is a simple measure of spread that gives the range covered by the middle half of the data; this distance is called the **interquartile range**. The interquartile range is useful primarily as a description of symmetric distributions; however, the individual quartiles (and median) are more informative in the case of a skewed distribution, because they portray the unequal spread of two sides of the distribution.

The calculation of the quartiles leads to another effective representation of a distribution, the **boxplot** or **box and whisker plot**. In a boxplot (Figure 4.6):

1 The ends of the box are at the quartiles, so that the length of the box is the interquartile range.

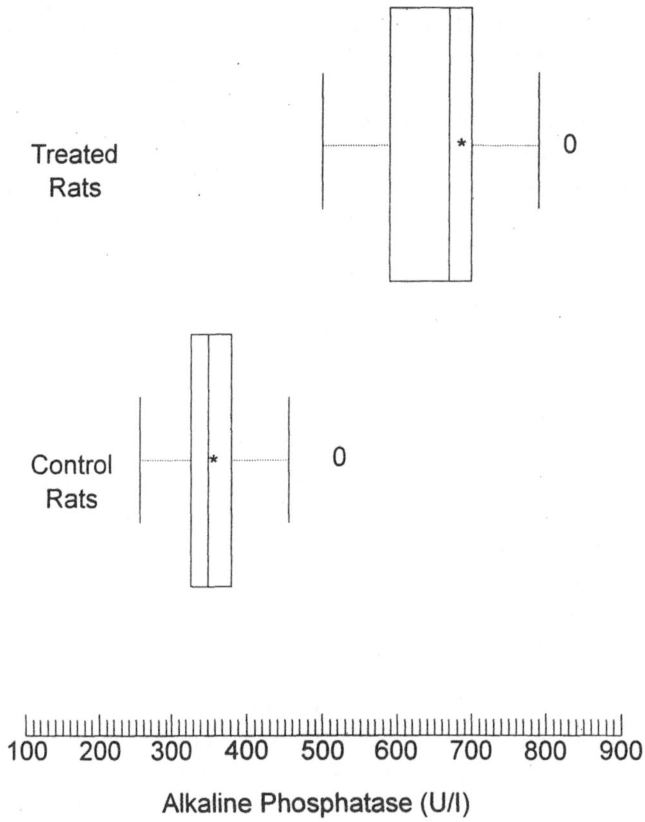

Figure 4.6 Box and whisker plots of alkaline phosphatase comparing treated and control rats

2 The median is marked by a line within the box.
3 The mean is marked by an asterisk.
4 The central vertical lines, called whiskers, extend from the box as far as the data extend, to a distance of at most 1.5 interquartile ranges. Any value more extreme is marked with a '0'.

There are several modifications of the above procedure but the general concept is the same in all cases, the centre, spread and overall range of the distribution are all immediately apparent.

The most commonly used measure of variability of a distribution is the **standard deviation**. This is a measure of spread about the mean, and should be used only when the mean is employed as a measure of centre. The **variance** of n observations $x_1, x_2, \ldots, x_n$ is

$$s^2 = \frac{\Sigma(x_i - \bar{x})^2}{n-1}$$

The standard deviation s is the square root of the variance s^2.

The idea behind the variance and the standard deviation as measures of spread is as follows. The deviations $x_i - \bar{x}$ display the spread of the values x_i about their mean $\bar{x}$. Some of these deviations will be positive and some negative because some of the observations fall on each side of the mean. So it is not possible to add the deviations to get an overall measure of spread, in fact the sum of the deviations will be zero. Squaring the deviations makes them all positive, and observations far from the mean in either direction will have large positive squared deviations. The variance is roughly the average squared deviation, but we divide by $n-1$ instead of n. This is due to the sum of the deviations always being zero, thus the last deviation can be found once we know the first $n-1$. Therefore the average is found by dividing by $n-1$, as we are not averaging by n unrelated numbers as only $n-1$ of the squared deviations can vary freely. The variance, s^2, will be large if the observations are widely spread about their mean, and small if the observations are all close to the mean. As an example, if we use the alkaline phosphatase data from the boxplot example:

365, 368, 335, 330, 414, 359, 356, 329, 304, 322, 357, 260, 395, 453, 406, 514, 353, 329, 257, 353.

The mean of this distribution is:

$$\bar{x} = \frac{365 + 368 + \ldots + 353}{20} = 357.95$$

To compute the variance we subtract 357.95($\bar{x}$) from each observation, square the resulting deviations, add, and divide by 19 ($n-1$) as follows:

$$s^2 = \frac{(365 - 357.95)^2 + (368 - 357.95)^2 + \ldots + (353 - 357.95)^2}{19} = 3546.68$$

The standard deviation is therefore:

$$s = \sqrt{3546.68} = 59.55$$

The task of calculating s^2 by hand is tedious. Many calculators will compute s or s^2 directly from keyed-in data, with no intermediate work on your part.

4.3 Some Basic Considerations in Experimental Design

The design and analysis of experiments is an extensive subject to which numerous books have been entirely devoted (Davies, 1954; Cochran and Cox, 1957). The problems of design are of course inseparable from those of analysis and it is worth emphasizing that unless a suitable design is employed, it may be very difficult or even impossible to obtain valid conclusions from the resulting data.

Statistical experimental design was founded in the early 1920s by R. A. Fisher, working at an agricultural research station at Rothamsted, England, where work such as comparing the yield from several varieties of wheat was performed. However, so many other factors were involved, such as temperature, humidity, position in field and so on, that unless the experiment was carefully designed, it would have been impossible to separate the effects of the different varieties of wheat from the effects of other variables. Fisher identified three important principles of experimental design: replication, randomization and blocking.

4.3.1 Replication

By **replication** we mean the application of each treatment of interest to more than one experimental unit (animal, panellist etc.). Replication has two benefits. Firstly, the population mean response for a particular treatment is estimated by averaging the observed responses across all the units receiving that treatment. The greater the number of individual observations contributing to such a treatment sample mean, the greater the precision with which the treatment population mean is estimated or, equivalently, the greater the power of the experiment in detecting differences between treatment population means.

Secondly, two units that receive the same treatment will not necessarily yield the same response, as variation in response between identically treated units is due to experimental error. By examining the variation in response between units within treatments, an estimate of the experimental error variance can be made. This forms a baseline against which all other apparent differences can be measured, in particular, differences between the various treatment sample means.

4.3.2 Randomization

The allocation of treatments to units should always be **randomized**. If not carried out, it is not possible to determine whether treatment differences are due to treatment or to confounding by other relevant factors. Randomization is practised to ensure that every treatment is equally likely to be advantaged or disadvantaged by the selection of units. It also serves to allow the scientist to proceed as if the assumption of independence is valid. That is, there is no available (known) systematic bias in how the data are obtained.

4.3.3 Blocking

If the individual units are homogeneous, respond in a consistent way to the various treatments, then the experimental error variance will be small. Therefore, the experiment will have high precision and it will be relatively easy to detect differences between the individual treatment population means. However, Fisher noticed that in agricultural research, the individual plots in a field were anything but homogeneous, they differed with respect to fertility, drainage etc. Furthermore, it was felt that such heterogeneity between units was desirable, in order to give the experiment wide coverage, so that it would produce results applicable to the real world. Thus, Fisher was faced with two apparently irreconcilable requirements – to design high-precision experiments using widely heterogeneous units.

Fisher introduced the idea of **blocking** to overcome this problem. Individual experimental units are grouped together into blocks in such a way that within each block, the units are as homogeneous as possible. Each block usually contains a single replicate of all the treatments, with the allocation of units to treatments within each block being randomized. In the subsequent analysis, the experimental error variance is estimated from units within the same block, so is relatively small, preserving the precision and power of the experiment. However, the blocks themselves may be as heterogeneous as the experimenter wishes, thus giving the experiment as wide a coverage as may be desired. For example, in a clinical trial it may be that a precise comparison could be effected by restricting the age, sex, clinical condition and other features of the patient, but these restrictions make it difficult to generalize the results. Therefore, the factors should be used as blocks.

4.3.4 *Statement of Objectives*

The goal of experimental design is statistical efficiency and the economizing of resources. Most experienced practitioners of experimental design have a sequence of steps to follow. The first is that the objective of the experiment should be clearly and fully stated. Having chosen the appropriate response variables, the factors which might influence them should be identified, and a decision taken as to which of these are to be studied in the current experiment. The statement of objectives should also include a description of the target population. The most common faults in setting out the objectives of an experiment are excessive vagueness about the hypotheses to be tested and the effects to be estimated, and excessive ambition!

4.3.5 *Sample Size*

An experiment is only worthwhile if it is powerful enough to detect differences between the various treatment population means of a magnitude that the scientist deems to be important. The ability of an experiment to discriminate depends upon:

1 The critical difference (the size of the effect it is desired to detect).
2 The significance level (the probability of an effect being detected when none exists).
3 The power (the probability of correctly detecting the difference as significant).
4 The design itself (how well it controls experimental error).
5 The number of replicates tested (the number of units per treatment).

Precise details of how to decide upon the appropriate size of experiment are discussed in a variety of books, and the reader is referred to Desu (1990). However, it is worth stressing that an over-precise experiment is just as much to be avoided as an under-precise one. An experiment which can detect differences between treatment population means, which are too small to be of any practical (as opposed to statistical) significance is wasteful of valuable resources that could otherwise have been put to better use.

Determining sample size is a compromise between available resources, power, expected variability of the outcome measure and effect size. The first two on this list are usually known or specified, but the latter two are not. When you have obtained as much information as possible (How important is the question? Has it been tackled before? etc.), the statistician will then come up with some values. However, this should not be seen as final, but rather as an opening bid in a bargaining procedure. It is not a one-way process, but rather an iterative one. At the end of a consultation, the experimenter should have a range of options; the width of this range will be proportional to the uncertainty about the various options.

In designing experiments, we should allow for involuntary censoring on sample size. In other words, a study might start off with enough units for analysis, but provide no margin of error should any unit be withdrawn before the end of the experiment. Just enough experimental units per group frequently leaves too few at the end to allow meaningful statistical analysis, and allowances should be made accordingly in establishing group size.

4.4 Statistical Analysis

It is important to realize from the outset that the observations are usually a **sample** from the set of all possible outcomes of the experiment (the **population**). A sample is taken because it is too expensive and time-consuming to take all possible measurements. Statistics are based on the idea that the sample will be 'typical' in some way and that it will enable us to make predictions about the whole population.

The data usually consist of a series of measurements on some feature of an experimental situation or on some property of an object. The phenomenon being investigated is usually called the **variate**. It is useful to distinguish between types of observation:

1 **Nominal**, measurements at various unordered discrete levels, examples are blood type, hair colour.
2 **Ordinal**, measurements at various ordered discrete levels, examples are clinical observations on animals and severity of lesions.
3 **Continuous**, measurements which can assume a continuous uninterrupted range of values, examples are weight, blood pressure.

Strictly, each type of response requires a different sort of statistical technique. Methods for nominal data can be used for ordinal or continuous data by defining categories, but this is rather wasteful of data. Methods for ordinal data can, and perhaps should, be used where one is uncertain of the underlying distribution, for analysing continuous data.

Although various guidelines (US Food and Drug Administration, 1984) suggest that statistical tests be used to evaluate the data generated from toxicological studies, there are no current standard statistical procedures being used to analyse clinical pathological data (Waner, 1992). There are some published (Gad and Weil, 1987; Lee, 1993) recommended approaches which summarize the methods to be used. One possible approach for continuous data is shown in Figure 4.7. The majority of these methods can be found in most standard statistical texts (Gad and Weil, 1987; Armitage and Berry, 1988; Campbell, 1989).

The practice of statistics involves many numerical calculations, some quite simple and some very complex. As you learn how to perform these calculations remember that the goal of statistics is not calculation for its own sake, but gaining understanding from data. A thorough grasp of the principles of statistics will

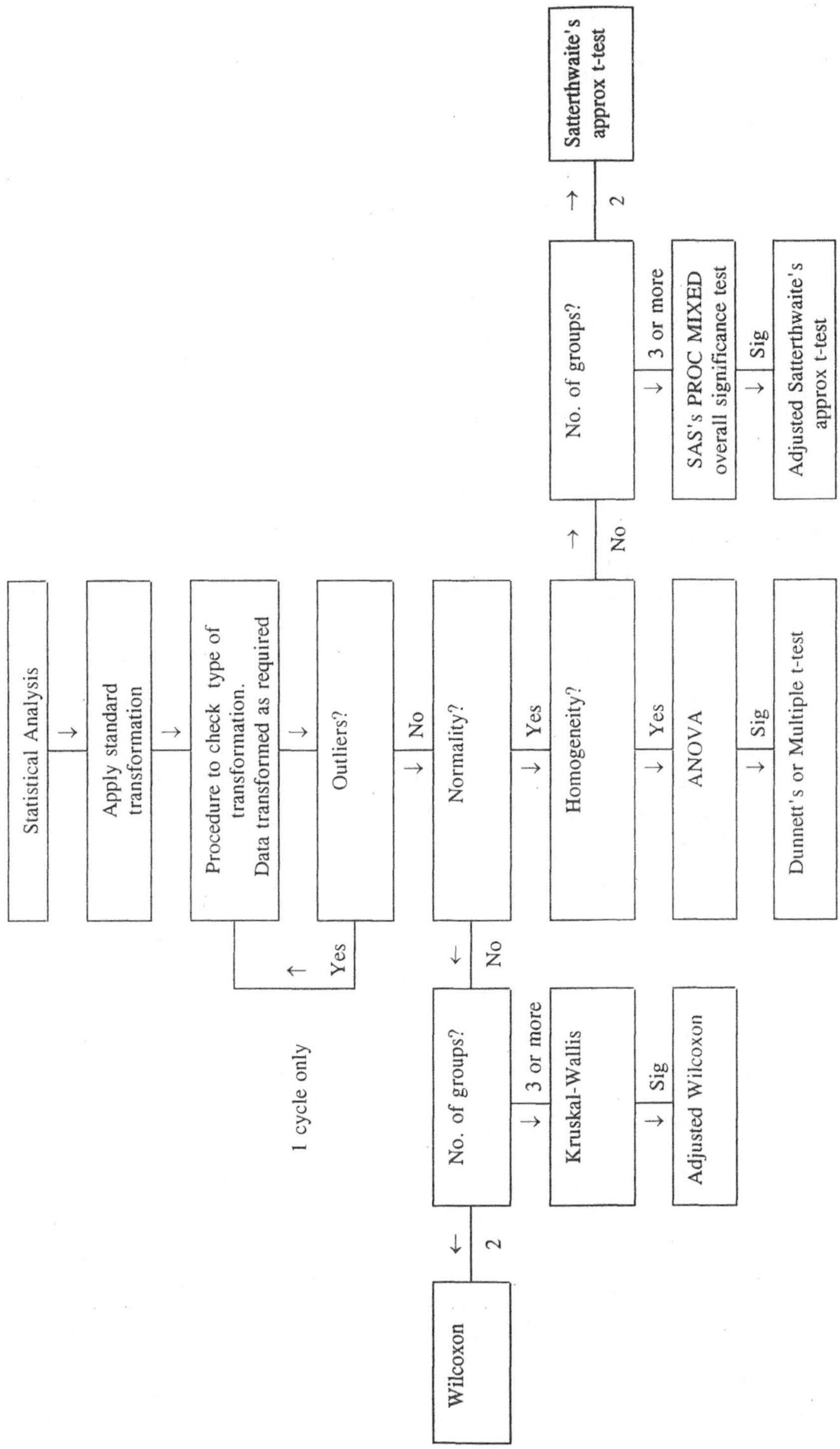

Figure 4.7 A simple approach for statistical analysis of clinical pathology data from a routine 28- or 90-day study

enable you to learn quickly more advanced methods as needed. On the other hand, a fancy computer analysis carried out without attention to basic principles will often produce elaborate nonsense. Thus the need to use the advice of an expert statistician cannot be overemphasized!

Suggested Reading

Basic Statistics

ARMITAGE, P. & BERRY, G. (1988) *Statistical Methods in Medical Research*, 2nd Edn. Oxford: Blackwell Scientific Publications.

CLARKE, G. M. & COOKE, D. (1992) *Basic Course in Statistics*, 3rd Edn. Kent: E. Arnold.

GAD, S. & WEIL, C. S. (1987) *Statistics and Experimental Design for Toxicologists*. New Jersey: Telford Press.

Intermediate Statistics

SNEDECOR, G. W. & COCHRAN, W. G. (1980) *Statistical Methods*, 7th Edn. Ames: Iowa State University Press.

WETHERILL, G. B. (1981) *Intermediate Statistical Methods*. London: Chapman & Hall.

Humorous Statistics

GONICK, L. & SMITH, W. (1993) *The Cartoon Guide to Statistics*. New York: HarperCollins.

References

ARMITAGE, P. & BERRY, G. (1988) *Statistical Methods in Medical Research*, 2nd Edn. Oxford: Blackwell Scientific Publications.

CAMPBELL, R. C. (1989) *Statistics for Biologists*, 3rd Edn. Cambridge: University Press.

COCHRAN, W. G. & COX, G. M. (1957) *Experimental Designs*, 2nd Edn. New York: John Wiley.

DAVIES, O. L. (1954) *The Design and Analysis of Industrial Experiments*. London: Oliver & Boyd.

DESU, M. M. (1990) *Sample Size Methodology*. London: Academic Press.

GAD, S. & WEIL, C. S. (1987) *Statistics and Experimental Design for Toxicologists*. New Jersey: Telford Press.

HUFF, D. (1954) *How to Lie with Statistics*. London: Gollancz.

LEE, P. N. (1993) *Statistics for Toxicology*. In Ballantyne, B., Marrs, T. & Turner, P. (eds) *General and Applied Toxicology*, Vol. 1, pp. 39–48. Basingstoke: The Macmillan Press.

TUKEY, J. W. (1977) *Exploratory Data Analysis*. London: Addison-Wesley.

US FOOD AND DRUG ADMINISTRATION (1984) *Department of Health and Human Services: Good Laboratory Practice Regulations*; proposed rule. Fed. Reg. 49: 43530–43537.

WANER, T. (1992) *Current Statistical Approaches to Clinical Pathological Data from Toxicological Studies*, **20**, 477–9.

5

General Enzymology

G. O. EVANS

Enzymes are proteins which have catalytic properties that include specific activation of their respective substrates. Much emphasis is placed on the application of plasma enzymes as markers of organ damage, with many enzymes used in toxicological studies to measure cellular injury, enzyme induction, activation or inhibition of enzymes. The distribution of enzymes in different tissues varies between species, and therefore influences their diagnostic value in particular species (Clampitt and Hart, 1978; Keller, 1981; Lindena *et al.*, 1986; Milne and Doxey, 1987; Davy *et al.*, 1988). The tissue distribution of an enzyme can also be affected by age and sex, and it may vary in the different cell types within an organ, e.g. kidney and liver (Braun *et al.*, 1983). Caution must be exercised when interpreting data on tissue enzyme distribution when a single extraction and assay system has been used, as this does not allow for the variable requirements for optimizing enzyme extraction and for the measurement of the enzyme.

The intracellular distribution of enzymes also varies and the proportions may be such that an enzyme can be regarded as relatively specific to a particular type of organelle. Several enzymes are cytosolic, for example lactate dehydrogenase, whilst other enzymes are located in organelles, such as glutamate dehydrogenase in mitochondria or acid phosphatase in lysosomes. Some enzymes such as aspartate aminotransferase occur both in mitochondria and in the cytosol, whereas other enzymes may be largely membrane bound, e.g. gamma glutamyl transferase.

Many enzymes show molecular heterogeneity, and the different forms are termed isoenzymes (or isozymes). Isoenzymes catalyse the same reaction and occur in the same species, but they differ in physical and chemical properties and can therefore be differentiated using electrophoretic, chromatographic, immunological, selective inhibition or modified substrate methods. Currently, the term isoenzyme should be restricted to the enzyme forms that originate at the level of the genes which encode the structure of the particular enzyme. The isoenzyme patterns in different tissues vary and their measurement may help in identifying the source of changes in total enzyme activity. There are also some non-genetic causes of multiple forms of enzymes (see isoforms in Chapter 10, Cardiotoxicity and Myotoxicity).

Table 5.1 Plasma enzymes and their respective half-lives in three species

Enzyme	Range of estimates for enzyme half-life (h)		
	Dog	Rat	Man
Aspartate aminotransferase	3.3 to 4.4	2.3	4 to 46
Alanine aminotransferase	2.5 to 60.9	4.4	32 to 52
Creatine kinase	0.6 to 16.2	0.6	3.5 to 35
Lactate dehydrogenase	1.6	–	40 to 114

The plasma activity of an enzyme depends on several factors including the enzyme concentrations in different tissues, the intracellular location of the enzyme, the rate of synthesis of the enzyme, the severity of tissue and cellular damage, the molecular size of the enzyme, and the rate of clearance of the enzyme from plasma. Given the tissue and cellular variations between species, it is not surprising to find marked differences in plasma enzymes between species (Lindena and Trautschold, 1986). For example, plasma alkaline phosphatase (ALP) and lactate dehydrogenase (LDH) activities show greater variability in rats and monkeys when compared to man, and therefore have a poorer predictive or diagnostic value in these species. In young animals the osseous ALP isoenzyme is the dominant plasma form in most species, but the proportion of intestinal ALP in plasma is far greater in the adult rat than in other species. Plasma gamma glutamyl transferase (GGT) activities are lower in most rodent species compared with the enzyme levels found in man.

It should not be assumed that enzymes applied in human clinical practice will show the same diagnostic value in the various animal species (Woodman, 1981). The majority of plasma enzyme measurements for detecting hepatotoxicity are not specific to the liver and show a wide tissue distribution. Several of these enzymes are also affected by damage to extra-hepatic tissue, for example aspartate aminotransferase and LDH following injury to cardiac or skeletal muscles. In the earlier stages of tissue damage, cytoplasmic enzymes may leak from cells where membrane permeability has altered. As the severity of tissue damage progresses, enzymes normally present in subcellular organelles will be released into the circulation (Schmidt and Schmidt, 1987; Bouma and Smit, 1988).

Plasma enzymes may also be classified as (i) plasma-specific enzymes, (ii) other secreted enzymes, e.g. amylase, and (iii) intracellular enzymes. The plasma-specific enzymes include those enzymes which are secreted by some organs and have a direct action in the plasma, for example, pseudo- or non-specific cholinesterase, and coagulation enzymes. With this class of enzyme, organ damage often correlates with reduced plasma enzyme activities, e.g. reduced pseudo-cholinesterase levels following some forms of hepatic injury.

Attention in toxicology studies is focused mainly on increased plasma activities following cellular injury. In some studies the plasma enzyme activity may show a good correlation with the extent of tissue injury but this is not always the case. Tissue regeneration may also be accompanied by changes in plasma enzymes. After enzyme leakage from the cell, plasma enzyme activities may fall but at different rates depending on the rates of clearance from the circulation. Often the published estimates for the clearances (or half-lives) of plasma enzymes are found to vary because of different methodologies used to establish these values. Table 5.1 shows some of the estimated elimination half-lives for enzymes, and it is particularly important to consider these variables in acute or time-course studies where enzyme changes occur.

Urinary enzymes may originate from:

1 proteins of low molecular weight passing into the glomerular filtrate
2 the renal tubular cells either by desquamation or injury
3 desquamation or injury of the epithelial cells of the urogenital tract
4 secretion by the glands of the urogenital tract
5 infiltrate or exudate cells, e.g. leucocytes
6 bacteria
7 erythrocytes
8 tumour cells (Raab, 1972).

The roles and interpretation of urinary enzymes are discussed in Chapter 7 on Nephrotoxicity.

Over 2000 enzymes have now been described and many were identified historically by using the substrate and by adding the suffix -ase, e.g. phosphat*ase*, to help with their classification. The International Union of Biochemistry uses a four-group Enzyme Commission number (EC) to define the enzyme by its function, and the enzymes are assigned to one of six classes:

1 oxidoreductases
2 transferases
3 hydrolases
4 lyases
5 isomerases
6 ligases

Table 5.2 shows some of the enzymes which have been used in toxicological studies.

5.1 Some of the Common Enzyme Measurements

5.1.1 Alanine aminotransferase (ALT)

ALT (also known as glutamic pyruvic transaminase, GPT) catalyses the reaction:

$$\text{L-alanine} + \text{alpha oxoglutarate} \rightleftharpoons \text{pyruvate} + \text{glutamate}$$

In many species, the proportion of ALT in hepatic tissue is greater than in any other organs, e.g. rat and dog, but in other species the proportions of ALT in hepatic and cardiac tissue are similar, e.g. rabbit (Clampitt and Hart, 1978;

Table 5.2 Enzymes, their abbreviations and Enzyme Commission (EC) numbers

Abbreviation	Recommended name	EC number
ALT(GPT)	alanine aminotransferase	2.6.1.2.
AAP	alanine aminopeptidase	3.4.1.2.
ALP	alkaline phosphatase	3.1.3.1.
AMY	amylase	3.2.1.1.
AST(GOT)	aspartate aminotransferase	2.6.1.1.
CHE	cholinesterase	3.1.1.8.
CK(CPK)	creatine kinase	2.7.3.2.
GGT	gamma glutamyl transferase	2.3.2.2.
GLDH	glutamate dehydrogenase	1.4.1.3.
α-HBD	α-hydroxybutyrate dehydrogenase	–
ICDH	isocitrate dehydrogenase	1.1.1.42.
LDH	lactate dehydrogenase	1.1.1.27.
LAAP	leucine arylamidase	3.4.11.2.
LAP	leucine aminopeptidase	3.4.11.1.
LIP	lipase	3.1.1.3.
NAG	N-acetylglucosaminidase	3.2.1.30.
5′NT	5′nucleotidase	3.1.3.5.
OCT	ornithine carbamoyl transferase	2.1.3.3.
SDH	sorbitol dehydrogenase	1.1.1.14.

Lindena *et al.*, 1986). General texts often state that ALT is a cytosolic enzyme, although mitochondrial ALT is present in many tissues of some species including the rat (DeRosa and Swick, 1975; Ruscak *et al.*, 1982). It is the relative proportions of cytosolic to mitochondrial forms that lead to such statements. Measurements of plasma ALT are not tissue specific and although ALT isoenzymes exist, they are not widely measured in diagnostic enzymology at present. In some species such as the common marmoset, plasma ALT activity is low and is less useful as a marker of hepatotoxicity (Davy *et al.*, 1984; Cowie and Evans, 1985).

5.1.2 *Aspartate Aminotransferase (AST)*

AST (also known as glutamic oxaloacetate transaminase, GOT) catalyses the reaction:

$$\text{aspartate} + \text{2-oxoglutarate} \rightleftharpoons \text{L-glutamate} + \text{oxaloacetate}$$

As for ALT, this enzyme is widely distributed in the tissues including cardiac, skeletal, hepatic and renal tissues. It is commonly used in conjunction with ALT to identify the site of tissue damage. Some common texts emphasize its use for non-hepatic tissues, but plasma levels of this enzyme do change following the administration of various hepatotoxins. Mitochondrial and cytosolic forms of AST exist and the proportion of mitochondrial to cytosolic form is generally greater than for ALT and thus may indicate the extent of cellular damage.

5.1.3 *Alkaline Phosphatases (ALP)*

This group of relatively non-specific enzymes hydrolyses a variety of ester orthophosphates under alkaline conditions, and exhibits optimal activity between pH 9 and 10. The group may act as hydrolases liberating inorganic phosphate or as phosphotransferases which transfer the liberated inorganic phosphate directly to an acceptor molecule such as a sugar. It is common practice to refer to the measurement as though it is a single enzyme, but the common analytical assays cover a variety of alkaline phosphatases. ALP is widely distributed in tissues, notably on the border membranes of the bile canaliculi and on sinusoidal surfaces of the liver, the intestinal mucosa, the osteoblasts of bone, the renal proximal tubules, the placenta and mammary glands. The cellular locations of ALP are generally on secretory or absorptive surfaces such as the brush border of the proximal convoluted tubules in the kidney, in the intestinal mucosa and in osteoblasts. In most species, age-related changes of osseous ALP are observed which reflect periods of bone growth.

Several plasma isoenzymes of alkaline phosphatase are recognized and these include hepatic, osseous, placental and intestinal forms. The relative proportions of these isoenzymes in plasma vary with species; for example in the rat, the osseous and intestinal isoenzymes are the dominant forms in plasma whereas the hepatic form is the dominant isoenzyme in plasma of other animals such as the dog. Thus food intake has a marked effect on plasma ALP values in the rat reflecting the intestinal component, but this does not occur with other laboratory animals (Madsen and Tuba, 1952; Sukumaran and Bloom, 1953; Pickering and Pickering 1978a, 1978b). In addition in the dog, the administration of glucocorticoids induces another form of ALP, so-called 'steroid-induced ALP' (Dorner *et al.*, 1974; Eckersall, 1986).

Many methods have been described for the separation and quantification of the ALP isoenzymes as an increase in total plasma ALP activity is difficult to interpret given the widespread distribution of this enzyme. These methods include various forms of electrophoretic separation, heat lability, and use of isoenzyme selective inhibitors such as urea and levamisole (Moss, 1982).

5.1.4 *Aminopeptidases and Arylamidases*

These enzyme assays have been linked here as they have some common applications in diagnostic enzymology.

Alanine aminopeptidase (AAP) and *leucine arylamidase* (LAAP) hydrolyse peptides at the N-terminal amino acid and some amino acid amides; the enzymes respectively hydrolyse leucyl- and alanyl-4-nitroanilide substrates. They occur in microsomes and are also membrane bound, and they have been used in studies of both hepatotoxicity and nephrotoxicity. These two enzymes should not be confused with cytosolic *leucine aminopeptidase* (LAP): this enzyme is an aminopeptidase which hydrolyses N-amino acid residues of proteins, in particular those with an N-terminal 1-leucine where 1-leucyl-β-napthylamide is commonly used as substrate.

5.1.5 *Amylase and Lipase (AMY and LIP)*

These enzymes are considered together as they are used primarily in evaluating changes in the pancreas. These two low-molecular-weight enzymes are found mainly in the pancreas although amylase is also found in salivary glands and it may be useful in monitoring changes in these glands. (See Chapter 9, Gastrointestinal Toxicity.)

5.1.6 *Creatine Kinase (CK)*

(Also known as creatine phosphokinase, CPK.) This enzyme is used primarily in studies of cardiotoxicity and myotoxicity and will be discussed more fully in Chapter 10.

5.1.7 *Pseudo- (or Non-specific) Cholinesterase and Acetylcholinesterase (CHE)*

Whilst of some interest in hepatic disease, these enzymes have much greater importance when dealing with some neurotoxic agents (see Chapter 11).

5.1.8 *Glutamate Dehydrogenase (GLDH or GDH)*

This is essentially a mitochondrial enzyme occurring in liver, kidney and muscle tissues and the highest concentrations are present in the liver. The enzyme catalyses the conversion of glutamate to 2-oxoglutarate, and plasma levels are generally low in most species.

5.1.9 *Gamma Glutamyl Transferase (GGT)*

(Also called gamma glutamyl transpeptidase.) This enzyme catalyses the transfer of gamma glutamyl groups from glutamyl peptides to an acceptor peptide or L-amino acid. The highest concentrations of GGT are found in the kidney, pancreas and then liver. In the rat kidney the level of GGT is approximately 200 times higher than the level found in hepatic tissue. The enzyme is located in the brush border cells of the renal proximal convoluted tubules and on the canicular surfaces of the hepatic parenchymal cells. Despite its relatively high tissue concentrations in the kidney, plasma GGT does not appear to alter following renal injury but urinary GGT measurements are helpful in monitoring renal tubular damage.

In hepatic studies, plasma GGT can be used as an indicator of cholestasis even in rats where plasma GGT levels are normally very low, often less than 2 IU/L. The use of GGT as a marker of enzyme induction and of the presence of hepatic tumours is less predictive in laboratory animals compared with data from human studies (Braun *et al.*, 1987; Batt *et al.*, 1992; see Chapter 6).

5.1.10 *Lactate Dehydrogenase (LDH)*

This catalyses the reversible oxidation of lactate to pyruvate and it is a cytosolic enzyme. It is distributed widely in the tissues and the tissue isoenzyme distribution patterns vary from species to species (Cornish *et al.*, 1970; Karlsson and Larsson, 1971; Milne and Doxey, 1987) (see Chapter 10 – Cardiotoxicity and Myotoxicity for further discussion). In mice, the lactate dehydrogenase toga-virus may affect plasma LDH values (Hayashi *et al.*, 1988, 1992).

5.1.11 *5′-Nucleotidase (5′NT)*

This enzyme is an alkaline phosphomonoesterase which catalyses the hydrolysis of nucleoside 5′monophosphates, e.g. adenosine-5′-monophosphate and inosine-5′-monophosphate. It appears to be distributed widely in the body tissues, and to be mainly a membrane-bound enzyme, which is useful in studies of hepatobiliary injury.

5.1.12 *Ornithine Carbamoyl Transferase (OCT)*

This catalyses the reaction:

$$\text{carbamoyl phosphate} + \text{ornithine} \rightleftharpoons \text{citrulline} + \text{phosphate}$$

The high concentration of OCT in hepatic tissue relative to other tissues leads to it being regarded as a 'liver-specific' enzyme, however its usefulness is sometimes limited by the technical requirements for the assay (Weingand *et al.*, 1992; Evans, 1993). This enzyme measurement should not be confused with ornithine decarboxylase (ODC) which is of interest in studies of polyamine metabolism (Carakostas, 1988; Evans, 1989a).

5.1.13 *Sorbitol Dehydrogenase (SDH)*

(Also known as iditol dehydrogenase.) This enzyme catalyses the interconversion of DL-sorbitol to fructose. General use has been confined largely to human medicine, but it can be useful in canine and non-human primate studies as a confirmatory test for hepatotoxicity since the enzyme distribution is confined mainly to the liver.

5.1.14 *Other Enzymes*

Arginase, malate dehydrogenase (MDH), isocitrate dehydrogenase (ICDH), glucose-6-phosphatase dehydrogenase (G6PDH) and alcohol dehydrogenase (ADH) are amongst the many enzymes which have been measured in studies of hepatotoxicity.

Glutathione transferases (GST), essential in many detoxification processes, have until recently not been used widely due to the low levels in plasma, poor enzymatic stability and inhibition by bilirubin and bile acids in the assay. The sensitive radioimmunoassays now available have been used in a few laboratories, but newer enzyme immunoassays may widen the use of this enzyme assay.

5.2 Enzyme Measurements

Enzyme activities are expressed generally as International Units where one unit is the amount of enzyme that will catalyse under optimum conditions the transformation of 1 μmol of substrate per minute. The activities are expressed as IU/L or mIU/ml. The katal has been proposed as an alternative unit, and it is defined as the catalytic activity which under defined conditions converts 1 mole of substrate per second thus the units are mol/s, and it may be expressed by the symbol kat (or nanokatal, nKat). One IU/L corresponds to 16.67 nKat/L. Unfortunately these definitions of enzyme units do not include other major variables which affect their measurement such as identity and pH of buffer, identity and concentration of substrate, temperature and the presence of activators. This has led to a variety of national and international recommendations for the common enzyme measurements which confuse comparisons between published data.

Enzymes may be measured by monitoring either a reduction in substrate concentration or an increase of reaction product. The majority of the common enzymes can be measured using a kinetic approach where the velocity of the enzyme reaction is monitored by serial measurements. For some enzymes the measured reactions are linked to the enzyme cofactors NAD or NADP, and changes in these cofactors are measured in the ultraviolet spectrum. Reactions can be linked through second or third step reactions where these cofactors are involved, e.g. AST and ALT. Other enzymes may be measured colorimetrically, e.g. ALP and GGT.

Enzyme reactions accelerate with increased temperature and the Q_{10} for the majority of enzyme reactions varies from 1.7 to 2.5, i.e. the reaction approximately doubles for every 10°C rise. Similarly, several enzymes are thermally inactivated, e.g. at temperatures of 56°C and above the activity of some ALP isoenzymes is reduced. Conversely, urinary GGT is deactivated when frozen. Although plasma CK is inactivated rapidly on storage, incubation with a high redox potential thiol, such as N-acetyl cysteine, will overcome this problem.

Dooley (1979) reported that substrate concentrations required for measurements of AST differed substantially between rat, dog, monkey and human sera. Reagents described as having optimal or optimized reagent concentrations for human plasma are not optimal for laboratory animal samples. Nevertheless, most laboratories routinely use such reagents formulated for use with human samples primarily for convenience and in the absence of data which show that these reagents are totally inappropriate for use with differing species used in toxicology. Variations between species using different enzyme substrates have been shown for cholinesterase (Myers, 1953; Evans, 1990) and other enzymes, e.g. angiotensin converting enzyme (Evans, 1989b). For alkaline phosphatase, the majority of methods employ 4-nitrophenylphosphate as the substrate, but there are two main alternative buffers – diethanolamine and 2-aminopropanol – which can cause interspecies differences (Masson and Holmgren, 1992).

For isoenzymes, the majority of laboratories currently use a variety of electrophoretic separation methods. Selective inhibition of isoenzymes with antibodies is being used increasingly, but there are problems associated with protein specificity and relative isoenzyme concentrations in animal samples.

Preanalytical factors discussed in an earlier chapter also need to be considered when measuring enzymes. Sites of sampling, particularly from the retro-orbital plexus in rodents, have marked effects on enzymes (Neptun *et al.*, 1985; Izumi *et al.*, 1993). For several enzyme measurements it is preferable to use plasma rather than serum because of the release of erythrocytic enzymes during clotting processes (Korsrud and Trick, 1973; Friedel and Mattenheimer, 1976). Some enzymes are present at relatively high concentrations in erythrocytes compared to plasma and therefore may interfere with the measurements, for example adenylate kinase in CK assays. The presence of haemolysis in plasma or serum samples affects enzyme values (Czerwek and Bleuel, 1981) and platelets also can increase apparent plasma CK and LDH values (Shibata and Kobayashi, 1978; Evans, 1985).

As for many other parameters, animals may show a high degree of intra-animal variability for enzyme measurements, e.g. ALT and AST in rats (Waner *et al.*, 1991), CK in dogs (Davies, 1992). Biorhythms have been described for urinary enzymes (Grotsch *et al.*, 1985; Do *et al.*, 1992). Age and sex differences occur for several enzymes, e.g. sex difference for cholinesterase in rats, urinary N-acetylglucosaminidase in beagles (Nakamura *et al.*, 1983) and age differences for urinary GGT in Wistar rats (Stoykova *et al.*, 1983).

Not all enzyme changes can be explained by the variables already mentioned. For example, plasma ALT also exhibits other changes which may complicate interpretation; a 50% food restriction over 210 days in rats resulted in elevated values compared to controls (Schwartz *et al.*, 1973), whilst Kast and Nishikawa (1981), in a study of fasting on the oral toxicity of several xenobiotics, reported reduced levels of ALT in the fasted groups compared to controls.

Again, when using plasma ALT where the main emphasis is on detecting increased enzyme activity, there are now several examples where the enzyme values fall due to effects on pyridoxal phosphate, which is a cofactor necessary for the action of both the aminotransferases AST and ALT (Dhami *et al.*, 1979; Waner and Nyska, 1991). Such effects may confuse the interpretation of data where hepatotoxicity occurs leading to increases of plasma ALT. Further complications with ALT have been described by Wells and To (1986) in covalent binding studies with acetoaminophen.

Some investigators have expressed the activities of the two aminotransferases enzymes as a ratio (AST:ALT) to assist with interpretation, but it is necessary for each laboratory to establish its own discriminating ratios as the use of different methods affects the values obtained for these ratios. From the factors affecting enzyme measurements in different species described here, it can be seen that ratios obtained for one species cannot be used for any other species.

In summary, several enzymes should be chosen with differing sensitivity and specificity for the detection of major organ toxicity. The choice of these enzymes should reflect their different tissue and intracellular locations. In more detailed studies, sampling times should be used with some reference to enzyme production and clearance related to the toxic insult. The use of such approaches is exemplified in the following chapters on nephrotoxicity, hepatotoxicity and cardiotoxicity. Whether through lack of technical simplicity or suitability for use with automated laboratory analysers, the number of enzymes measured in regulatory studies has not increased dramatically over the previous two decades.

References

BATT, A. M., SIEST, G., MAGDALOU, J. & GALTEAU, M.-M. (1992) Enzyme induction by drugs and toxins. *Clinica Chimica Acta*, **209**, 109–21.

BOUMA, J. M. W. & SMIT, M. J. (1988) Elimination of enzymes from plasma in the rat. In Goldberg, D. M., Moss, D. W., Schmidt, E. & Schmidt, F. W. (eds), *Enzymes – Tools and Targets, Advances in Clinical Enzymology*, Vol. 6, pp. 111–19. Basel: Karger.

BRAUN, J. P., BERNARD, P., BURGET, V. & RICO, A. G. (1983) Tissue basis for use of enzymes in toxicology. *Veterinary Research and Communications*, **7**, 331–5.

BRAUN, J. P., SIEST, G. & RICO, A. G. (1987) Uses of γ-glutamyltransferase in experimental toxicology. In Cornelius, C. E. & Rico, A. G. (eds) *Advances in Veterinary Science and Comparative Medicine*, Vol. 31, pp. 151–72. London: Academic Press.

CARAKOSTAS, M. D. (1988) What is serum ornithine decarboxylase? *Clinical Chemistry*, **34**, 2606–7.

CLAMPITT, R. B. & HART, R. J. (1978) The tissue activities of some diagnostic enzymes in ten mammalian species, *Journal of Comparative Pathology*, **88**, 607–21.

CORNISH, H. H., BARTH, M. L. & DODSON, V. N. (1970) Isozyme profiles and protein patterns in specific organ damage. *Toxicology and Applied Pharmacology*, **16**, 411–23.

COWIE, J. R. & EVANS, G. O. (1985) Plasma aminotransferase measurements in the marmoset (*Callithrix jacchus*). *Laboratory Animals*, **19**, 48–50.

CZERWEK, H. & BLEUEL, H. (1981) Normal values of alkaline phosphatase, glutamic oxaloacetic transaminase and glutamic pyruvic transaminase in the serum of experimental animals using optimised methods and the effect of haemolysis on these values. *Experimental Pathology*, **19**, 161–3.

DAVIES, D. T. (1992) Enzymology in preclinical safety evaluation. *Toxicologic Pathology*, **20**, 501–5.

DAVY, C. W., BROCK, A., WALKER, J. M. & EICHLER, D. A. (1988) Tissue activities of enzymes of diagnostic interest in the marmoset and rat. *Journal of Comparative Pathology*, **99**, 41–53.

DAVY, C. W., JACKSON, M. R. & WALKER, J. M. (1984) Reference intervals for some clinical chemistry parameters in the marmoset (*Callithrix jacchus*): effect of age and sex. *Laboratory Animals*, **18**, 135–42.

DEROSA, G. & SWICK, R. W. (1975) Metabolic implications of the distribution of the alanine aminotransferase isoenzymes. *Journal of Biological Chemistry*, **250**, 7961–7.

DHAMI, M. S. I., GRANGOVA, R., FARKAS, R., BALOZS, T. & FEUER, G. (1979) Decreased aminotransferase activity of serum and various tissue in the rat after cefazolin treatment. *Clinical Chemistry*, **25**, 1263–6.

DO, T.-X., BOISNARD, P., GIRAULT, A., PLANCHENAULT, P., BREGET, R. & PRELOT, M. (1992) Etude expérimentale des variations diurnes et nocturnes de quelques activités enzymatiques rénales de rats Sprague–Dawley. *Science et techniques de l'animal de laboratoire*, **17**, 207–11.

DOOLEY, J. F. (1979) The role of clinical chemistry in chemical and drug safety evaluation by use of laboratory animals. *Clinical Chemistry*, **25**, 345–7.

DORNER, J. L., HOFFMAN, W. E. & LONG, G. B. (1974) Corticosteroid induction of an isoenzyme of alkaline phosphatase in the dog. *American Journal of Veterinary Research*, **35**, 1457–8.

ECKERSALL, P. D. (1986) Steroid induced alkaline phosphatase in the dog. *Israel Journal of Veterinary Medicine*, **42**, 253–9.

EVANS, G. O. (1985) Lactate dehydrogenase activity in platelets and plasma. *Clinical Chemistry*, **31**, 165.

(1989a) More on orthinine decarboxylase. *Clinical Chemistry*, **35**, 897–8.

(1989b) Species relationships for plasma angiotensin converting enzyme activity using a furanacryloyl tripeptide substrate. *Experimental Animals*, **38**, 897–8.

(1990) Species relationships for plasma acylcholine-acyl hydrolase using three different substrates. In *Fourth Congress International Society for Animal Clinical Biochemistry*, p. 270. Davis, USA: University of California.

(1993) Clinical pathology testing recommendations for nonclinical toxicity and safety studies. *Toxicologic Pathology*, **21**, 513–14.

FRIEDEL, F. & MATTENHEIMER, H. (1976) Release of metabolic enzymes from platelets during blood clotting of man, dog, rabbit and rat. *Clinica Chimica Acta*, **30**, 152–9.

GROTSCH, H., HROPOT, M., KLAUS, E., MALERCZYK, V. & MATTENHEIMER, H. (1985) Enzymuria of the rat: biorhythms and sex differences. *Journal of Clinical Chemistry and Clinical Biochemistry*, **23**, 343–7.

HAYASHI, T., OZAKI, M., MORI, M., SAITO, MITOH, T. & YAMAMOTO, H. E. (1992) Enhanced clearance of lactic dehydrogenase-5 in severe combined immunodeficiency (SCID) mice: effect of lactic dehydrogenase virus on enzyme clearance. *International Journal of Experimental Pathology*, **73**, 173–81.

HAYASHI, T., SALATA, K., KINGMAN, A. & NOTKINS, A. L. (1988) Regulation of enzyme levels in the blood. *American Journal of Pathology*, **132**, 503–11.

IZUMI, Y., SUGIYAMA, F., SUGIYAMA, Y. & YAGAMI, K.-I. (1993) Comparison between the blood taken from orbital plexus and heart in analyzing plasma bichemical values – increase of plasma enzyme values in the blood from orbital sinus. *Experimental Animals*, **42**, 99–102.

KARLSSON, B. W. & LARSSON, G. B. (1971) Lactic and malic dehydrogenases and their multiple forms in the mongolian gerbil as compared with the rat, mouse and rabbit. *Comparative Biochemistry and Physiology*, **40B**, 93–108.

KAST, A. & NISHIKAWA, J. (1981) The effects of fasting on oral acute toxicity of drugs in rats and mice. *Laboratory Animals*, **15**, 359–64.

KELLER, P. (1981) Enzyme activities in the dog: tissue analyses, plasma values, and intracellular distribution, *American Journal of Veterinary Research*, **42**, 575–82.

KORSRUD, G. O. & TRICK, K. D. (1973) Activities of several enzymes in serum and heparinised plasma from rats. *Clinica Chimica Acta*, **48**, 311–15.

LINDENA, J., SOMMERFELD, U., HOPFEL, C. & TRAUTSCHOLD, I. (1986) Catalytic enzyme activity concentration in tissues of man, dog, rabbit, guinea pig, rat and mouse. *Journal of Clinical Chemistry and Clinical Biochemistry*, **24**, 35–47.

LINDENA, J. & TRAUTSCHOLD, I. (1986) Catalytic enzyme activity concentration in plasma of man, sheep, dog, cat, rabbit, guinea pig, rat and mouse. *Journal of Clinical Chemistry and Clinical Biochemistry*, **24**, 11–18.

MADSEN, N. B. & TUBA, J. (1952) On the source of alkaline phosphatase in rat serum. *Journal of Biological Chemistry*, **195**, 741–50.

MASSON, P. & HOLMGREN, J. (1992) Comparative study of alkaline phosphatase in human and animal samples using methods based on AMP and DEA buffers: effect on quality control. *Scandinavian Journal of Clinical and Laboratory Investigations*, **52**, 773–5.

MILNE, E. M. & DOXEY, D. L. (1987) Lactate dehydrogenase and its isoenzymes in the tissues and sera of clinically normal dogs. *Research in Veterinary Science*, **43**, 222–4.

MOSS, D. W. (1982) Alkaline phosphatase isoenzymes. *Clinical Chemistry*, **28**, 2007–16.

MYERS, D. K. (1953) Studies on cholinesterase: 9. Species variation in the specificity pattern of the pseudocholinesterases. *Biochemical Journal*, **55**, 67–79.

NAKAMURA, M., ITOH, T., MIYATA, K., HIGASHIYAMA, N., TAKESUE, H. & NISHIYAMA, S. (1983) Difference in urinary N-acetyl-β-D-glucosaminidase activity between male and female beagle dogs. *Renal Physiology (Basel)*, **6**, 130–3.

NEPTUN, D. A., SMITH, C. N. & IRONS, R. D. (1985) Effect of sampling site and collection method on variations in baseline pathology parameters in Fischer-344 rats. I. Clinical chemistry. *Fundamental and Applied Toxicology*, **5**, 1180–5.

Pickering, C. E. & Pickering, R. G. (1978a) Studies of rat alkaline phosphatase. I. Development of methods for detecting isoenzymes. *Archives of Toxicology*, **39**, 249–66.

Pickering, R. G. & Pickering, C. E. (1978b) Studies of rat alkaline phosphatase. II. Some implications of the methods for detecting the isoenzymes of plasma alkaline phosphatase in rats. *Archives of Toxicology*, **39**, 267–87.

Raab, W. P. (1972) Diagnostic value of urinary enzyme determinations. *Clinical Chemistry*, **18**, 5–25.

Ruscak, M., Orlicky, J. & Zubor, V. (1982) Isoelectric focussing of the alanine aminotransferase isoenzymes from the brain, liver and kidney. *Comparative Biochemistry and Physiology*, **71B**, 141–4.

Schmidt, E. & Schmidt, F. W. (1987) Enzyme release. *Journal of Clinical Chemistry and Clinical Biochemistry*, **25**, 525–40.

Schwartz, E., Tornaben, J. A. & Boxhill, G. C. (1973) The effects of food restriction on hematology, clinical chemistry and pathology in the albino rat. *Toxicology and Applied Pharmacology*, **25**, 515–24.

Shibata, S. & Kobayashi, B. (1978) Blood platelets as a possible source of creatine kinase in rat plasma and serum. *Thrombosis and Haemostasis (Stuttgart)*, **39**, 701–6.

Stoykova, S., Phillipon, C., Labaille, F., Prevot, D. & Manuel, Y. (1983) Comparative study of alanine-amino-peptidase and gamma-glutamyl-transferase activity in normal Wistar rat urine. *Laboratory Animals*, **17**, 246–51.

Sukumaran, M. & Bloom, W. L. (1953) Influence of diet on serum alkaline phosphatase in rats and men. *Proceedings of the Society for Experimental Biology and Medicine*, **84**, 631–4.

Waner, T. & Nyska, A. (1991) The toxicological significance of decreased activities of blood alanine and aspartate aminotransferase. *Veterinary Research Communications*, **15**, 73–8.

Waner, T., Nyska, A. & Chen, R. (1991) Population distribution profiles of the activities of blood alanine and aspartate aminotransferase in the normal F344 inbred rat by age and sex. *Laboratory Animals*, **25**, 263–71.

Weingand, K., Bloom, J., Carakostas, M., Hall, R., Helfrich, M., Latimer, K., Levine, B., Neptun, D., Rebar, A., Stitzel, K. & Troup, C. (1992) Clinical pathology testing recommendations for nonclinical toxicity and safety studies. *Toxicologic Pathology*, **20**, 539–43.

Wells, P. G. & To, E. C. A. (1986) Urine acetaminophen hepatoxicity: temporal interanimal variability in plasma glutamic-pyruvic transaminase profiles and relation to *in vivo* chemical covalent binding. *Fundamental and Applied Toxicology*, **7**, 17–25.

Woodman, D. D. (1981) Plasma enzymes in drug toxicity. In Gorrod, J. W. (ed.), *Aspects of Drug Toxicity Testing Methods*, pp. 145–56. London: Taylor & Francis.

6

Assessment of Hepatotoxicity

D. D. WOODMAN

6.1 Liver Anatomy and Physiology

The liver is composed of angular, generally hexagonal structures called lobules. In the centre of each lobule is a central vein, which is a branch of the hepatic vein and along the sides of the lobule are branches of the hepatic artery, the portal vein, bile ducts and lymph vessels. Both the portal vein and hepatic artery drain into the central vein via specialized capillaries called sinusoids. Although the lobule is the standard anatomical unit, it does not constitute a microcirculatory unit. A functional circulatory unit was described by Rappaport in 1954 and is called an acinus (Rappaport *et al.*, 1954). The hepatic acinus is a group of cells arranged around a branch of the portal vein with its associated hepatic artery and bile duct, which drains across the acinus into the central veins of two adjacent lobules (Rappaport, 1976). The acinus is also divided into three microcirculatory zones. Zone 1 is the closest to the portal vein and thus constitutes the periportal area of the lobule. This zone receives blood rich in oxygen and nutrients, the concentrations of which fall as blood drains across the acinus to the central vein. Zones 2 and 3 are thus more prone to anoxia and toxic damage. Zone 3 roughly constitutes the centrilobular area of the lobule. This circulatory zoning appears also to be allied with some metabolic differentiation of the acinar zones.

The sinusoids are lined by specialized endothelial cells, a type of fixed macrophage called Kupffer cells, which are phagocytic and form part of the reticuloendothelial system. The main purpose of these cells is to remove particulate material. Outside the Kupffer cells is a layer of parenchymal cells or hepatocytes. These are large cuboidal cells which make up the majority of the parenchymal volume. The space between the Kupffer cells and the parenchymal cells is called the space of Disse, where interstitial tissue fluid forms and drains back to the lymph vessels at the edge of the lobule.

Between the parenchymal cells are small bile canaliculi which carry bile formed in the cell to the bile ducts at the edge of the lobule. Anatomically then, the structure of the liver is designed to carry materials to the parenchymal cell for processing and to remove the products of its action to a variety of destinations.

Within the cell, the mitochondria are responsible for much of the diverse metabolic activity of the liver. Ribosomes are responsible for protein synthesis and the Golgi apparatus is involved in storage, transportation and secretion of many formed components.

Liver damage can conveniently be divided into two main categories: liver cell damage, where direct hepatocyte damage or destruction occurs, and impairment of bile flow, where the normal flow of bile is reduced without necessarily involving hepatocyte destruction. Within these two groups, however, many different patterns of damage can be identified, depending on the specificity of the original damage.

In considering what tests are appropriate for assessing liver function or damage it is important to appreciate the wide variety of processes with which the liver is involved. It is also vital to remember that functional deficit and damage are not the same and may occur quite independently, especially in the early stages of toxicity.

6.2 Functions of the Liver

With the exception of certain specialized proteins such as the immunoglobulins, most plasma proteins are synthesized by the liver. Protein is also catabolized by the liver to form urea as an end product, which is then excreted via the kidneys.

Carbohydrates can be synthesized by the liver from fats and proteins and ATP can be synthesized via the Embden–Meyerhof pathway using glucose. The liver is a major store of carbohydrate, which is stored as glycogen, and it plays a major role in the maintenance of blood glucose by glycogenolysis and gluconeogenesis. The liver not only uses fats to produce carbohydrates, it also synthesizes a number of lipids such as cholesterol and triglycerides, which it can also modify to form other complex lipids and lipoproteins. In addition to these major metabolic activities, it synthesizes a number of complex organic structures such as haem, which is incorporated into haemoglobin for normal oxygen transport, and cytochrome P450, a class of detoxifying enzymes. The liver even synthesizes insulin-like growth factors under stimulation by growth hormone, and as such qualifies as an endocrine organ.

The other major role of the liver is its central action in detoxification and excretion. While the kidney is largely responsible for the excretion of water-soluble materials, the liver metabolizes lipid-soluble materials to produce molecules which are more water soluble and less toxic. These processed, unwanted materials can then either be excreted directly via the bile to be eliminated from the intestine, or if sufficiently soluble can be excreted by the kidney.

The liver, then, is capable of a wide range of activities, and this is reflected by the number of diagnostic tests reflecting these actions which can be used to detect changes in liver function or damage.

6.3 Tests of Hepatic Function and Damage

Many tests may be used to investigate damage to or functional aspects of the liver, and most have been derived from human clinical medicine. Usefulness in humans does not guarantee similar usefulness in animals, but many have been successfully adapted and established in animal clinical diagnostics. Some are more effective than others, and it is these tests which in general are the most frequently used, though there are a large number of tests which are used occasionally or under special conditions.

6.3.1 Bilirubin

Bilirubin is the breakdown product of haem, the porphyrin part of the haemoglobin molecule. It is highly insoluble and toxic and it is particularly useful diagnostically because the liver is responsible both for its conjugation with glucuronic acid, rendering it more soluble, and for its excretion. It is most common to measure the total bilirubin in plasma, but the conjugated and unconjugated forms, formerly known as direct and indirect bilirubin respectively, can be measured separately and may be diagnostically useful.

In man, an increase in plasma bilirubin manifests itself clinically as jaundice at concentrations above approximately 70 μmol/L. Cholestasis may result in plasma levels significantly in excess of 340 μmol/L. The increase in the yellow pigmentation of separated plasma becomes obvious at about 30 μmol/L, i.e. less than twice the upper normal limit.

With the exception of genetically defined strains such as Gunn rats, both rats and dogs have much lower plasma bilirubin levels. The upper limit of normal is of the order of 9 μmol/L or less. Rosenthal *et al.* (1981) using HPLC suggest that actual values in rats are 0.3–1.5 μmol/L, and are thus at or below the limit of accurate detection by routinely used photometric methods. Dogs in particular have a very low renal threshold for conjugated bilirubin (Mills and Dragstedt, 1938), and as a result, plasma levels of the order of 50 μmol/L are generally indicative of hepatocellular diseases. Compared with man, small increases in plasma bilirubin are significant indicators of hepatic damage in animals, and increases in urine bilirubin occur earlier in animals than in man.

The measurement of total and conjugated bilirubin for the differential diagnosis of obstructive and haemolytic jaundice can be applied to animals as well as man. However, due to the lower initial concentration of plasma bilirubin in animals, this approach is of limited use.

The low concentration of bilirubin in the plasma of small animals is the result of the low renal threshold for conjugated bilirubin, but clearance of free bilirubin is much less efficient. Consequently, haemolytic jaundice in animals can result in much higher plasma bilirubin than that which occurs with obstructive jaundice. Kernicterus rarely develops in small animals (Tryphonas and Rozdilsky, 1970) since the hepatic bilirubin conjugating system develops before birth. However, compared with man, the binding capacity of albumin is less, so in severe cases of unconjugated hyperbilirubinaemia, precipitation of bilirubin can occur due to saturation of the albumin binding sites. In general, non-human primates are similar to man, although the rhesus monkey can concentrate bilirubin in bile to a much greater extent (Gartner *et al.*, 1971).

Because of the low renal threshold in dogs, conjugated bilirubin is detectable in urine at a very early stage, in fact often before the development of bilirubinaemia (Gardiner and Parr, 1967). This can, however, be confusing as slightly increased levels of bilirubin may be evident in any febrile state. Unconjugated bilirubin bound to albumin does not readily pass through the glomerulus, therefore bilirubinaemia should be a good indicator of a haemolytic condition. Unfortunately, when haemolysis results in saturation of the haptoglobins, bilirubin appears in the urine before there is any significant increase in plasma conjugated bilirubin (De Schepper and van der Stock, 1971, 1972) suggesting that some bilirubin formation and conjugation takes place in the dog kidney.

6.3.2 Bile Acids

Unlike bilirubin, bile acids are not waste products, but are formed in the liver from cholesterol and secreted into the intestine to act as lipid emulsifiers. The measurement of bile acids has historically presented great problems, but the advent of radioimmunoassays and the availability of enzymatic methods have allowed the routine use of bile acid measurement to become a tenable proposition. In most instances of human clinical practice, plasma bilirubin or alkaline phosphatase has been the measurement of choice in confirming obstructive jaundice. Once clinical jaundice has developed, the high sensitivity of bile acids is of limited use, serving only to confirm the obvious.

In the sphere of toxicology, the sensitivity of bile acid measurement presents a positive advantage. Due to difficulties with bilirubin and alkaline phosphatase measurements in laboratory animals, bile acid measurement is an attractive indicator of cholestatic episodes (Gopinath *et al.*, 1980; Woodman and Maile, 1981; Hauge and Abdelkader, 1984). Hepatic transport of bile acids involves processes similar to those for bilirubin, i.e. uptake, conjugation and biliary excretion, but the routes are not shared with bilirubin. Hepatic uptake of bile acids is carrier mediated, concentration dependent (Erlanger *et al.*, 1976) and different from that by which bilirubin or exogenous dyes are handled (Scharschmidt *et al.*, 1975). A single pass through the liver is sufficient to remove the bulk of bile acids present (Hofmann, 1976). Under similar conditions only about 5% of bilirubin is removed while bromosulphthalein and indocyanine green fall between these two extremes (Martin *et al.*, 1975). This highly efficient process results in low circulating levels, but hepatocellular damage is reflected in increasing plasma bile acid levels much more quickly than would be the case for bilirubin.

The transport system involved in biliary secretion of bile acids is also different from that of bilirubin (Alpert *et al.*, 1969). Furthermore, bile salts and acids are largely reabsorbed from the intestine. In hepatobiliary disorders the efficient bile salt removal from plasma will again result in an early increase in plasma levels if excretory efficiency is impaired. The increased intestinal absorption of bile acids following a meal has been used in man to enhance the sensitivity of this index. In the two hours following a meal, the accelerated enterohepatic circulation amplifies minor deficiencies in the hepatic transport system which result in reduced secretion of bile acids into the bile. This procedure is not routinely practical in experimental animals.

Analytically, the measurement of total 3-hydroxy bile acids is simple and can effectively be applied to all species. Although this is less sensitive than measurement of individual bile acids, species differences in bile acid profiles detract from the use of individual bile acid measurements as a general approach (Parraga and Kaneko, 1985). The rat in particular is unusual in that whereas cholic acid is the major primary bile acid in most species, the rat has a predominance of the secondary bile acid, β-muricholic acid which increases rapidly during cholestasis. Man does not possess the appropriate enzyme systems involved in the formation of β-muricholic acid and this bile acid is not found in human plasma (Hofmann, 1988).

6.3.3 *Dye Excretion Tests*

The excretory capacity of the liver can be studied by following the excretion of exogenous dyestuffs such as bromosulphthalein (BSP, sulphobromothalein) indocyanine green (ICG) or rose bengal.

Despite the fact that bromosulphthalein clearance (BSP) was first used as a test of liver function in 1925 (Rosenthal and White, 1925) it remains one of the most sensitive tests available if properly conducted. Other dyes have also been described, but BSP remains the most widely used. When administered intravenously, BSP binds principally to albumin (Baker and Bradley, 1966) and to a lesser extent lipoprotein (Baker, 1966). It is removed from the circulation almost exclusively by the hepatocytes where it is bound to glutathione-S-transferase and conjugated with glutathione (Jablonski and Owen, 1969). The resulting conjugate, together with a proportion of the free dye, is then excreted into the bile.

Both hepatic uptake and biliary excretion are saturable carrier-mediated processes shared with bilirubin and many other exogenously administered agents. Care must be taken to ensure that this competition is reduced to a minimum so that the ability of the liver to clear BSP in isolation is accurately measured.

Despite the undoubted potential of dye clearance and excretion tests, there are a number of disadvantages which ensure that they are only used infrequently in either a clinical or an experimental situation. Clearance tests using exogenous compounds are time consuming and require greater care to conduct and interpret. Infusion of the dye, timing of blood or urine sampling and accurate calculation of the rate of dye removal are not simple, and relatively small errors can have disproportionate effects on the final result. Administration of any exogenous substances can additionally pose a health hazard to the subject, and severe systemic reactions have been reported following BSP administration to man (Whitby *et al.*, 1984). These tests will remain part of the repertoire of tests of liver functional integrity, but their associated difficulties will always limit their practical application.

6.3.4 *Enzymes*

The measurement of enzyme activity in plasma provides a means of studying changes occurring in many tissues and organs. However, a large number of factors relating to the plasma activity of each enzyme must be carefully evaluated (Woodman, 1981; see also Chapter 5). Plasma enzymes usually offer an indirect reflection of tissue damage, appearing in the plasma at increased activities following their loss from damaged cells. The magnitude of the change will largely be a reflection of the intensity and duration of leakage, the intracellular location of a particular enzyme and its tissue distribution. The rate of synthesis and plasma clearance may also combine to produce a highly complex and often confusing picture. In particular, the wide tissue distribution of most enzymes makes organ specificity a difficult goal.

However, the advantage that enzyme measurement offers, which outweighs all of these potential disadvantages, is that of sensitivity. Because of the large difference between intracellular and plasma concentrations, plasma levels can rise rapidly as a result of enzyme loss from a relatively small number of cells. In the early stages of tissue damage, cytoplasmic enzymes may leak from cells whose membrane permeability has been increased by changes in cellular function. At this stage, enzymes associated with subcellular organelles will be unable to escape and will therefore not increase in the plasma. If the damage process continues and becomes more extensive throughout the cell, mitochondrial and microsomal enzymes will be released and plasma levels will rise. Such plasma changes will be among the first measurable alterations following liver damage.

This phased release can also act as an indicator of the time and severity of the original damage. Plasma half-lives of intracellular enzymes are short, usually no more than a few days at most, so a rapid rise following damage will quickly be cleared unless further damage releases more enzymes. This makes plasma enzyme measurements particularly useful in acute studies but severely limits their use in chronic studies where damage induced by initial dosing may regress with prolonged dosing.

It cannot be assumed that because an enzyme has been shown to be useful in human clinical practice it will be useful when dealing with laboratory animals. Care must be exercised in the choice of the enzymes used for a particular species when such a large number of candidate enzymes are available (Woodman, 1988).

Hepatocellular damage

The aminotransferases, aspartate and alanine aminotransferase (AST and ALT), were among the first enzymes to become established as diagnostic markers in man and animals. In most species ALT is an effective marker of liver damage and is widely used, particularly in conjunction with AST, to aid the differential diagnosis of liver damage (Sarker, 1974; Keller, 1981).

Dehydrogenases

Dehydrogenases have also emerged as useful indicators of liver damage. Lactate dehydrogenase (LDH) is widely used in human clinical diagnostic enzymology, but its wide tissue distribution in all species limits its use to a confirmatory role or as part of an enzyme pattern. The measurement of LDH in toxicity studies is now generally not recommended. Its usefulness can be greatly increased by isoenzyme differentiation for specific tissues, but this is not a routinely practical approach. Sorbitol dehydrogenase (SDH) which has shown a limited usefulness in man, has been shown to be largely liver derived in most animal species. Its measurement is one of the more specific indicators of hepatocellular damage (Dooley *et al.*, 1979; Abdelkader and Hauge, 1986).

Glutamate dehydrogenase (GDH) and isocitrate dehydrogenase (ICDH) show similar release patterns to SDH and although not as widely used, are effective markers of hepatocellular damage (Boyd, 1962; Keller, 1981). Malate dehydrogenase (MDH), while less commonly used, has also been applied to the diagnosis of hepatocellular damage (Zimmerman, 1982).

Urea cycle enzymes

Two of the urea cycle enzymes, ornithine carbamoyl transferase (OCT) and arginase can provide evidence of liver damage in many animal species (Cornelius *et al.*, 1963). Both enzymes offer organ specificity since the urea cycle is confined to the liver; however, their use has been limited by methodological difficulties, and they are not widely used (Carakostas *et al.*, 1986).

Cholestasis

Alkaline phosphatase (ALP) has long been the standard enzyme marker of cholestasis, despite its shortcomings. While ALP has been applied to most species, the low activity in the rat and cat liver, the high intestinal component in rat plasma and the variability in primate plasma activities do not make it an ideal choice. Isoenzyme separations can greatly improve the predictive specificity, but the required effort precludes this approach as a routine procedure (Kominami *et al.*, 1984). Alternative enzymes have also been used, but none has shown wide advantages over ALP. 5′Nucleotidase (5′NT) has been used as a more specific liver variant of ALP but with limited success. Leucine aminopeptidase (LAP) may have advantages in primates but not in other species (Kominami *et al.*, 1984).

Although there are many conflicting views in the literature, gamma glutamyl transpeptidase (GGT) has probably proved the most widely used alternative/adjunct to ALP, although in the dog, rat and cat activity in the liver itself is negligible. However, in rat, GGT has been claimed to be a specific indicator of bile duct lesions (Leonard *et al.*, 1984). GGT has also been claimed to be a useful indicator of hepatic tumours, as has phosphohexose isomerase (PHI), but neither has become established.

Measurements of other enzymes have been reported to be of use in the diagnosis of liver damage, but few have gained wide acceptance (Zimmerman, 1982). This highlights the fact that no single enzyme is ideal. The choice of enzymes used is based ultimately as much on the availability of background data and experience as on intrinsic diagnostic potential. The accurate diagnosis of hepatobiliary damage is greatly enhanced by the use of a battery of enzyme measurements. Unfortunately no group of enzyme tests offers the 'best' compromise. Species variation and advances in methodology, particularly with regard to isoenzymes, can influence which enzymes are chosen. However, while the specific enzyme choice may vary, a panel of at least four enzymes should always be included when assessing possible liver damage.

6.3.5 **Plasma Proteins**

Most plasma proteins can act as indicators of the synthetic capacity of the liver. Albumin, fibrinogen, α_1 antitrypsin, haptoglobin, caeruloplasmin, transferrin and prothrombin are all synthesized in hepatocytes. Following cellular damage the capacity to synthesize protein is reduced, and as the extent of damage increases, the levels of these proteins in the plasma will tend to decrease. As plasma protein half-lives are much longer than enzymes, the rate of synthesis required to maintain normal plasma levels is much lower. Decreases in plasma proteins therefore tend to reflect chronic damage. The measurement of individual proteins is usually accomplished by immunological methods and requires the preparation of specific antibodies. Albumin is easily measured using dye binding techniques (see Chapter 14).

Synthesis of acute-phase proteins, which include fibrinogen, complement and α_1 antitrypsin, is stimulated by tissue injury or inflammation (Koj, 1974). It appears that inflammation causes the release of IL-1 from macrophages in particular, which in turn stimulates the liver to synthesize and secrete acute-phase proteins (Sipe, 1985). Liver damage which is minor or in its early stages may paradoxically stimulate this synthesis while reducing overall synthetic capacity, initially maintaining normal total protein concentrations.

The common pattern seen following significant hepatocellular damage is a reduction in albumin accompanied by a relative increase in gamma globulins (which are synthesized by the B cells of the lymphoid system), often with little change in the level of total protein. These changes, however, produce an obvious effect on the albumin/globulin ratio and on the plasma protein electrophoretogram. The protein measurements most frequently used in the routine evaluation of liver function are, therefore, total protein, albumin, the calculation of albumin/globulin ratio and the plasma protein electrophoretic pattern.

6.3.6 Lipids

The liver is a major site of cholesterol synthesis and conversion into bile acids. During cholestatic episodes, cholesterol synthesis is increased, resulting in increased plasma concentrations of cholesterol and the low and very low density lipoproteins which act as transport proteins (see Chapter 13). Triglycerides, non-esterified fatty acids and phospholipids are also often increased in conjunction with the rise in cholesterol (Davison and Wills, 1974; Clampitt, 1978). Changes in plasma lipids are normally preceded by more obvious changes in other diagnostic tests, and only become significant when the dysfunction has developed to a serious extent. Additionally, changes in plasma lipids can be produced by factors other than liver damage and as a first-line diagnostic approach their usefulness is accordingly limited. It is the case, however, that liver damage often leads to fat accumulation in the hepatocyte (Lombardi, 1966) and plasma lipid measurements may help provide a better awareness of the potential development of fatty liver.

6.3.7 Drug-metabolizing Enzymes

The liver contains a number of drug-metabolizing systems, normally associated with the microsomes of the hepatocyte. A number of these enzymes, particularly the cytochrome P450 family, can be increased by induction. Compounds such as phenobarbitone increase the enzymes responsible for their metabolism. They may increase proliferation of the smooth endoplasmic reticulum and raise the concentration of specific fatty acids and bound phospholipids, and may cause a marked increase in liver weight. The response to different compounds is not uniform and different patterns of isoenzyme induction may be seen (Timbrell, 1993).

The clearance of administered antipyrene has been used as a measure of this drug-metabolizing activity either by measuring the decline of plasma antipyrene (HPLC), or by administration of ^{14}C aminopyrene. This undergoes microsomal demethylation to release ^{14}C formaldehyde which is finally metabolized to $^{14}CO_2$. The release of this radioactive gas can be measured in expired breath. This is a complicated test and basically a research tool.

Greater interest is being shown in drug-metabolizing enzymes and in their induction and decrease. These are not plasma measurements, however, and require determination in liver tissue. A variety of demethylating enzymes can be measured using this approach as can glucuronidating enzymes such as uridine diphosphate glucuronyl transferase and the mixed function oxidase system cytochrome P450 (Hinton and Grasso, 1993). These measurements are not easy because of the use of tissue rather than plasma, but they are *direct* measurements of liver metabolic capacity and may be particularly useful in pharmaceutical development.

6.3.8 Drug-induced Hepatotoxicity

The majority of drugs are taken orally and following absorption rapidly pass through the liver. This emphasizes the importance of the liver in metabolism and detoxification, but it also ensures that the liver is particularly vulnerable to toxic chemicals, and about 10% of adverse drug reactions involve the liver (Davis and Williams, 1977). Prior to 1950, the main toxicological interest in chemical pathology centred on the effects of the organic solvents such as chloroform and carbon tetrachloride and on elements such as arsenic, selenium and phosphorus. Very few compounds used therapeutically were even considered. However, the enhanced sensitivity of detection methods, together with the vast expansion in the number of compounds available for therapeutic use, has resulted in far greater attention being given to the detection of hepatic injury with this latter group of compounds.

Most of the originally studied toxic chemicals produced an acute hepatic necrosis, and carbon tetrachloride remains a model for this type of liver damage. The effect of this chemical is so drastic that following administration, the severe hepatocellular damage results in a plasma pattern of liver enzymes very similar to that in liver tissue. This pattern is also quite distinct from that produced by acute hepatitis where significant cellular damage also occurs, and the magnitude of the increase is also much greater following poisoning. In contrast to the acute hepatic poisons, whose effects are highly predictable, most of the drug-induced hepatic reactions are uncommon, idiosyncratic and often unrelated to dose level, though when a response is noted it is of a characteristic type.

Some of the drugs now known to have hepatotoxic side-effects produce acute

Table 6.1 Enzyme activities and bile acid concentrations in the rat following thioacetamide (TA) administration[a]

Treatment group	ALT	AST	GDH	SDH	Glycocholate	Cholate
	Activities, IU/L at 30°C					
Control	39.0(4.2)	55.9(4.2)	9.0(4.1)	27.7(4.2)	0.4(0.2)	1.1(0.7)
TA 20.0 mg/kg	51.0(5.9)***	68.5(8.7)***	22.9(10.6)***	86.7(37.7)***	1.5(0.6)***	3.1(1.9)**

[a]Arithmetic means (and SD) and statistical significances of difference from mean using student's t-test are shown. *$p<0.05$; **$p<0.01$; ***$p<0.001$.

Abbreviations: ALT–alanine aminotransferase; AST–aspartate aminotransferase; GDH–glutamate dehydrogenase; SDS–sorbitol dehydrogenase.

hepatic necrosis, e.g. tetracycline (Seto and Lepper, 1974) but more often a second type of reaction results in an intrahepatic cholestasis. This leads to a reduction or cessation of bile flow, with the resultant build-up of retained bilirubin, which may be visible clinically as jaundice. This response is produced by the phenothiazine derivatives, several antimicrobial agents, oral hypoglycaemics and anabolic steroids. There is also a group of compounds which can produce a massive necrosis and a serum enzyme pattern very similar to that seen in viral hepatitis. Many of the monoamine oxidase inhibitors fall into this category.

A vast number of hepatotoxins have been reported, and rather than attempt to compile a comprehensive list, it is much more useful to examine some individual hepatotoxins and their actions on the liver and the resultant plasma clinical chemistry. For more detailed reviews, see Zimmerman (1978) and Tucker (1982).

6.3.9 *Drug-induced Acute Hepatic Necrosis*

Organic solvents, such as chloroform and carbon tetrachloride, produce an acute hepatic necrosis, although at doses just below that required to induce overt necrosis, plasma enzyme levels, particularly SDH, ALT, AST and ICDH, are still significantly increased (Korsrud *et al.*, 1972).

Thioacetamide provides a good model of this type of damage, inducing a reproducible centrilobular necrosis in rats (Hunter *et al.*, 1977). Rats dosed at 20 mg/kg and killed 24 h later show a minimal hepatocyte necrosis visible by light microscopy together with a small increase in plasma enzymes and bile acids (Table 6.1).

This type of damage can be seen with common drugs such as paracetamol, where the damage is associated with P450-dependent enzyme conversion to a toxic metabolite (Dahlin *et al.*, 1984).

Many non-steroidal anti-inflammatory drugs can cause liver toxicity of various types, but most often this involves hepatocellular necrosis. Aspirin, indomethacin, ibufenac and fluproquazone represent examples of different classes of NSAIDs (non-steroidal anti-inflammatory drugs) which can cause hepatic necrosis, the latter two having been withdrawn from the market as a result (Lewis, 1984).

Dimethylnitrosamine administered chronically for 5–10 months to rats can cause liver tumours, but it is also a potent hepatotoxin in rats and dogs (Barnes and Magee, 1954). Chronic treatment with dimethylnitrosamine increases transaminase values in the dog and decreases indocyanine green clearance to about half the control level (Kawasaki *et al.*, 1984).

Carbon tetrachloride-induced hepatotoxicity has also been shown to affect both erythrocyte and hepatocyte plasma membranes. The activity of Na^+/K^+ and Ca^{2+} dependent ATPases decreased and cholesterol/phospholipid ratio rose significantly following treatment (Muriel *et al.*, 1992). In addition to necrosis, cytotoxic injury may also occur as fatty liver or steatosis. Both tetracycline and methotrexate can induce this type of damage (Timbrell, 1983). Tetracycline inhibits lipid transport and its effect in rats is dose dependent and rapid (Breen *et al.*, 1975). Despite this, the change in transaminase levels is generally less than that seen with drugs inducing necrosis.

6.3.10 Cholestasis

A second major area of drug-induced toxicity is cholestasis. A variety of drugs have been shown capable of causing cholestasis but in animals the effects are generally unpredictable. Oral administration of α naphthylisothiocyanate (ANIT) to rats, however, produces a highly predictable and dose-dependent cholestasis (Becker and Plaa, 1965). Cholestasis occurs 15–24 h after a single dose of ANIT in rats, mice and guinea pigs (Capizzo and Roberts, 1971) and damage to the cells lining the bile duct can be detected by electron microscopy 3 h after dosing (Schaffner *et al.*, 1973). The mechanism by which this damage occurs is not clear, and may well involve a number of different actions.

Experimentally, an increasing effect with dose can be seen on enzymes and bile acids in the rat. Doses of 60 and 120 mg/kg cause a marked increase in transaminases, GDH and ICDH, bilirubin and bile acids (Table 6.2a and b), with a slight degree of cholangitis visible by light microscopy. A dose of 30 mg/kg produces no changes detectable by light microscopy, but ALT and total cholate are still significantly increased (Woodman and Maile, 1981).

Chlorpromazine may produce a drug-induced cholestasis in up to 1% of patients taking the drug, producing a pattern of results similar to extrahepatic jaundice, i.e. increased plasma cholesterol, alkaline phosphatase and transaminases (Zimmerman, 1974). Similarly, a number of anabolic steroids such as methyltestosterone cause a cholestasis without any major parenchymal damage (Klatskin, 1975). Alkaline phosphatase and transaminases may only be slightly increased but bilirubin is often markedly elevated. In the dog, a steroid-inducible form of alkaline phosphatase has been reported which increases following steroid treatment without any accompanying liver damage (Eckersall, 1986).

6.3.11 Artefactual Changes

It is not unusual when developing novel chemical entities whose action is unknown, to record findings which reflect an interference with the analytical method rather than a real effect on the liver. Semiquantitative tests such as the ‘dipstick’ test for

Table 6.2a Enzyme activities in rat plasma after ANIT administration[a]

	AP	ALT	GDH	ICDH
Treatment group	Activities, IU/L at 30°C			
Control	587.3(86.8)	31.0(3.2)	7.9(3.1)	4.2(1.5)
ANIT, 30 mg/kg	576.2(106.0)	35.1(3.2)***	8.0(3.4)	3.5(2.5)
ANIT, 60 mg/kg	986.6(249.6)***	234.5(129.9)***	248.0(136.4)***	31.7(20.2)***
ANIT, 120 mg/kg	1285.8(193.8)***	439.9(129.3)***	397.9(160.8)***	78.0(29.4)***

Table 6.2b Bilirubin and bile acid concentrations in rat plasma after ANIT administration[a]

	Bilirubin	Glycocholate	Cholate	Chenodeoxy-cholate	Conjugated primary bile acids
Treatment group	Concentration, μmol/L				
Control	2.2(0.9)	1.6(1.0)	5.0(2.4)	1.1(0.7)	3.7(1.1)
ANIT, 30 mg/kg	3.4(0.9)**	2.4(1.0)*	10.1(3.6)***	1.8(0.9)*	4.6(1.7)
ANIT, 60 mg/kg	40.7(35.1)***	50.2(32.1)***	53.9(44.0)**	3.0(1.3)***	118.8(74.2)***
ANIT, 120 mg/kg	127.4(48.7)***	104.5(31.2)***	102.9(45.9)***	4.9(1.3)***	229.5(72.2)***

[a]Arithmetic means (and SD) and statistical significances of difference from mean using student's t-test are shown. *p<0.01; **p<0.01; ***p<0.001.

Abbreviations: AP–alkaline phosphatase; ALT–alanine aminotransferase; GDH–glutamate dehydrogenase; ICDH–isociatrate dehydrogenase.

urine bilirubin can produce an interaction with a chemical other than bilirubin which results in a colour on the stick which can be confused as a positive reaction. Though less common, effects on plasma enzymes have been reported when no adverse effect on the liver has been detectable. Although decreases in plasma enzymes are of limited diagnostic interest, the effect of cephazolin on aminotransferase activity provides an interesting example.

Administration of cephazolin to dogs results in a rapid decrease in plasma ALT (Dhami *et al.*, 1979). Figure 6.1 shows the time course of this reaction in two dogs following a single dose of 500 mg/kg of cephazolin. In native plasma, AST activity drops rapidly to levels which are almost undetectable 24 h after dose. These values gradually recover to predose levels after about three weeks. Treatment of native plasma with pyridoxal phosphate results in no loss in enzyme activity. The dramatic change in activity seen originally does not reflect any damage to the liver or any loss in its capacity to synthesize the enzyme, but an action directly on the enzyme molecule itself. Cephazolin interferes with the binding of pyridoxal phosphate, an essential coenzyme for aminotransferases, to the enzyme protein. Stripped of its coenzyme, the AST apoenzyme has no catalytic activity and is not measured by normal enzyme assay techniques, but treatment with excess pyridoxal phosphate restores the activity.

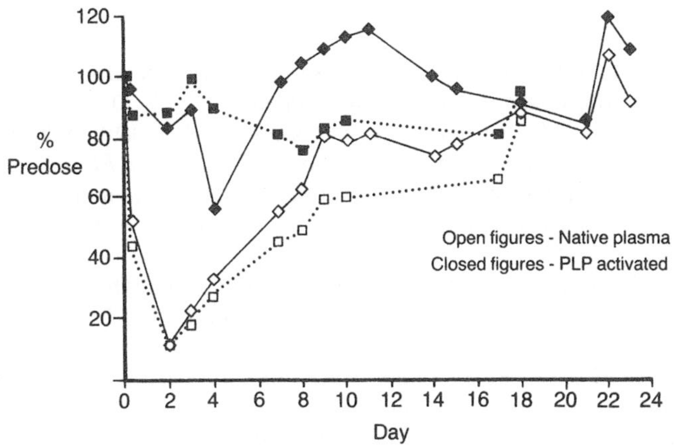

Figure 6.1 AST activity following cefazolin treatment

6.4 Conclusion

From the wide variety of possible liver function tests it is obvious that no single index offers an infallible warning of adverse occurrences. This is a result of the multifunctional role of the liver and of the relatively non-specific nature of most of the indices available. Realistic assessment of possible liver damage can only be made by a battery of tests chosen to cover the major functions and types of damage which can occur.

It would be quite wrong to attempt to provide a definitive battery of liver function tests. A more practical approach would be to adopt a phased strategy, including the parameters which are most often affected following liver damage. This approach would combine the most widely useful and sensitive tests suitable for the species under investigation, such as the appropriate cellular enzymes and bile acids combined with the less sensitive but still often useful tests such as alkaline phosphatase, bilirubin and albumin. Any abnormal results from such a test battery should then lead on to further appropriate tests which might help confirm the diagnosis.

The specific tests used would depend on the animal species under investigation and the local expertise of the laboratory involved, but careful choice of the most relevant tests can provide a highly sensitive means of detecting even minor damage. Improvements in analytical techniques and the introduction of new indices which supersede existing tests may modify this choice from time to time. Examples of such changes are the demise of the thymol turbidity test and the resurgence of interest in bile acids following the introduction of RIA and enzyme assays.

The importance of accurately assessing liver damage in toxicology demands this constant re-examination and improvement of the laboratory tests available. More work needs to be done in particular on the peripheral metabolism and excretion of released enzymes and on the metabolism and excretion of specific bile acids in different species. These areas offer some of the most promising possibilities for increasing the sensitivity of currently available tests and further clarifying the laboratory diagnosis of liver damage.

References

ABDELKADER, S. V. & HAUGE, J. G. (1986) Serum enzyme determination in the study of liver disease in dogs. *Acta Veterinaria Scandinavica*, **27**, 59–70.

ALPERT, E., MOSHER, M., SHANSKE, A. & ARIAS, I. M. (1969) Multiplicity of hepatic excretory mechanisms for organic anions. *Journal of General Physiology*, **53**, 238–47.

BAKER, K. J. (1966) Binding of sulphobromophthalein (BSP) and indocyanine green (ICG) by plasma α_1 lipoproteins. *Proceedings of the Society for Experimental Biology and Medicine*, **122**, 957–63.

BAKER, K. J. & BRADLEY, S. E. (1966) Binding of sulphobromophthalein (BSP) sodium by plasma albumin: its role in hepatic BSP extraction. *Journal of Clinical Investigation*, **45**, 281–7.

BARNES, J. M. & MAGEE, P. N. (1954) Some toxic properties of dimethylnitrosamine. *British Journal of Industrial Medicine*, **11**, 167–74.

BECKER, B. A. & PLAA, G. L. (1965) The nature of α-naphthylisothiocyanate-induced cholestasis. *Toxicology and Applied Pharmacology*, **7**, 680–5.

BOYD, J. W. (1962) The comparative activity of some enzymes in sheep cattle and rats – normal serum and tissue levels and changes during experimental liver necrosis. *Research in Veterinary Science*, **3**, 256–68.

BREEN, K. J., SCHENKER, S. & HEIMBERG, M. (1975) Fatty liver induced by tetracycline in the rat: dose–response relationship and effect of sex. *Gastroenterology*, **69**, 714–23.

CAPIZZO, F. & ROBERTS, R. J. (1971) α-Naphthylisothiocyanate (ANIT)-induced hepatotoxicity and disposition in various species. *Toxicology and Applied Pharmacology*, **19**, 176–87.

CARAKOSTAS, M. C., GOSSETT, K. A., CHURCH, G. E. & CLEGHORN, B. L. (1986) Evaluating toxin-induced hepatic injury in rats by laboratory results and discriminant analysis. *Veterinary Pathology*, **23**, 264–9.

CLAMPITT, R. B. (1978) An investigation into the value of some clinical biochemical tests in the detection of minimal changes in liver morphology and function in the rat. *Archives of Toxicology, Supplement*, **1**, 1–13.

CORNELIUS, C. E., DOUGLAS, G. M., GRONWALL, R. R. & FREEDLAND, R. A. (1963) Comparative studies on plasma arginase and transaminases in hepatic necrosis. *Cornell Veterinarian*, **53**, 181–91.

DAHLIN, D. C., MIWA, G. T., LU, A. Y. H. & NELSON, S. D. (1984) N-acetyl-p-benzoquinone imine: a cytochrome P-450-mediated oxidation product of acetaminophen. *Proceedings of the National Academy of Science USA*, **81**, 1327–31.

DAVIS, M. & WILLIAMS, R. (1977) Hepatic disorders. In Davies, D. M. (ed.), *Textbook of Adverse Drug Reactions*, pp. 146–72. Oxford: Oxford University Press.

DAVISON, S. C. & WILLS, D. (1974) Studies on the lipid composition of the rat liver endoplasmic reticulum after induction with phenobarbitone and 20-methylcholanthrene. *Biochemical Journal*, **140**, 461–8.

DE SCHEPPER, J. & VAN DER STOCK, J. (1971) Influence of sex on the urinary bilirubin excretion at increased free plasma haemoglobin levels in whole dogs and in isolated normothermic perfused dog kidneys. *Experientia*, **27**, 1264–5.

(1972) Increased urinary bilirubin excretion after elevated free plasma haemoglobin levels. 1. Variations in the calculated renal clearance of bilirubin in whole dogs. *Archives Internationales de Physiologie et de Biochimie*, **80**, 279–91.

DHAMI, M. S. I., DRANGOVA, R., FARKAS, R., BALAZS, T. & FEUER, G. (1979) Decreased aminotransferase activity of serum and various tissues in the rat after cefazolin treatment. *Clinical Chemistry*, **25**, 1263–6.

DOOLEY, J. F., TURNQUIST, L. J. & RACICH, L. (1979) Kinetic determination of serum sorbitol dehydrogenase activity with a centrifugal analyser. *Clinical Chemistry*, **25**, 2026–9.

Eckersall, P. D. (1986) Steroid induced alkaline phosphatase in the dog. *Israel Journal of Veterinary Medicine*, **42**, 253–9.

Erlanger, S., Glasinovic, J. C., Poupon, R. & Dumont, M. (1976) In Taylor, W. (ed.), *The Hepatobiliary System*, pp. 433–52. New York: Plenum Press.

Gardiner, M. R. & Parr, W. H. (1967) Pathogenesis of acute lupinosis of sheep. *Journal of Comparative Pathology*, **77**, 51–62.

Gartner, M. R., Lane, D. L. & Cornelius, C. E. (1971) Bilirubin transport by liver in adult *Macaca mulatta*. *American Journal of Physiology*, **220**, 1528–35.

Gopinath, C., Prentice, D. C., Street, A. E. & Crook, D. (1980) Serum bile acid concentration in some experimental liver lesions of rat. *Toxicology*, **15**, 113–27.

Hauge, J. G. & Abdelkader, S. V. (1984) Serum bile acids as an indicator of liver disease in dogs. *Acta Veterinaria Scandinavica*, **25**, 495–503.

Hinton, R. H. & Grasso, P. (1993) Hepatotoxicity. In Ballantyne, B., Marrs, T. & Turner, P. (eds), *General and Applied Toxicology*, Vol. 1, pp. 619–62. Basingstoke: Macmillan.

Hofmann, A. E. (1976) The enterohepatic circulation of bile acids in man. *Advances in Internal Medicine*, **21**, 501–34.

(1988) Bile acids. In Arias, I. M., Jakoby, W. B., Popper, M., Schachter, D. & Shafritz, D. A. (eds), *The Liver, Biology and Pathobiology*, pp. 553–72. New York: Raven Press.

Hunter, A. L., Holscher, H. A. & Neal, R. A. (1977) Thioacetamide-induced hepatic necrosis. I. Involvement of the mixed-function oxidase enzyme system. *Journal of Pharmacology and Experimental Therapeutics*, **200**, 439–48.

Jablonski, P. & Owen, J. A. (1969) The clinical chemistry of bromosulphophthalein and other choleophilic dyes. *Advances in Clinical Chemistry*, **12**, 309–89.

Kawasaki, S., Umekita, N., Beppu, T., Wada, T., Sugiyama, Y., Iga, T. & Hanano, M. (1984) Hepatic transport of indocyanine green in dogs chronically intoxicated with dimethylnitrosamine. *Toxicology and Applied Pharmacology*, **75**, 309–17.

Keller, P. (1981) Enzyme activities in the dog: tissue analyses, plasma values and intracellular distribution. *American Journal of Veterinary Research*, **42**, 575–82.

Klatskin, G. (1975) Toxic and drug-induced hepatitis. In Schiff, L. (ed.), *Diseases of the Liver*, 4th Edn, pp. 604–710. Philadelphia: J. P. Lippincot.

Koj, A. (1974) Acute phase reactants: their synthesis, turnover and biological significance. In Allison, A. C. (ed.), *Structure and Function of Plasma Proteins*, Vol. 1, pp. 73–131. London: Plenum Press.

Kominami, T., Oda, K. & Ikehara, Y. (1984) Induction of rat hepatic alkaline phosphatase and its appearance in serum: electrophoretic characterization of liver-membranous and serum soluble forms. *Journal of Biochemistry*, **96**, 901–11.

Korsrud, G. O., Grice, H. C. & McLaughlan, J. M. (1972) Sensitivity of several serum enzymes in detecting carbon-tetrachloride-induced liver damage in rats. *Toxicology and Applied Pharmacology*, **22**, 474–83.

Leonard, T. B., Neptun, D. A. & Popp, J. A. (1984) Serum gamma glutamyl transferase as a specific indicator of bile duct lesions in the rat liver. *American Journal of Pathology*, **116**, 262–9.

Lewis, J. H. (1984) Hepatic toxicity of nonsteroidal anti-inflammatory drugs. *Clinical Pharmacy*, **3**, 128–38.

Lombardi, B. (1966) Considerations on the pathogenesis of fatty liver. *Laboratory Investigation*, **15**, 1–20.

Martin, J. F., Mikulecky, M., Blaschke, T. F., Waggoner, J. G., Vergalla, J. & Berk, P. D. (1975) Differences between the indocyanine green disappearance rates of normal men and women. *Proceedings of the Society for Experimental Biology and Medicine*, **150**, 612–17.

Mills, M. A. & Dragstedt, C. A. (1938) Removal of intravenously injected bromosulphthalein from the bloodstream of the dog. *American Medical Association Archives of Internal Medicine*, **62**, 216–21.

MURIEL, P., FAVARI, L. & SOTO, C. (1992) Erythrocyte alterations correlate with CCl_4 and biliary obstruction induced liver damage in the rat. *Life Sciences*, **52**, 647–55.

PARRAGA, M. E. & KANEKO, J. J. (1985) Total serum bile acids and the bile acid profile as tests of liver function. *Veterinary Research Communication*, **9**, 79–88.

RAPPAPORT, A. M. (1976) The microcirculatory acinar concept of normal and pathological liver structure. *Beitrage zur Pathologie*, **157**, 215–43.

RAPPAPORT, A. M., BOROWRY, Z. J., LOUGHEED, W. M. & LOTTO, W. N. (1954) Subdivision of hexagonal liver lobules into a structural and functional unit: role in hepatic physiology and pathology. *Anatomical Record*, **119**, 11–34.

ROSENTHAL, P., BLANKAERT, N., KABRA, P. M. & THALER, M. M. (1981) Liquid chromatographic determination of bilirubin and its conjugates in rat serum and human amniotic fluid. *Clinical Chemistry*, **27**, 1704–7.

ROSENTHAL, S. M. & WHITE, E. C. (1925) Clinical application of the bromsulphthalein test for hepatic function. *Journal of the American Medical Association*, **84**, 1112–14.

SARKER, N. K. (1974) Alanine and aspartate aminotransferase activities in serum and various subcellular fractions from the livers of different species. *International Journal of Biochemistry*, **5**, 375–81.

SCHAFFNER, F., SCHARNBECK, H., HUTTERER, F., DENK, F., GREIM, H. A. & POPPER, H. (1973) Mechanisms of cholestasis. VII. α-Naphthylisothiocyanate-induced jaundice. *Laboratory Investigation*, **28**, 321–33.

SCHARSCHMIDT, B. F., WAGGONER, J. G. & BERK, P. D. (1975) Hepatic organic anion uptake in the rat. *Journal of Clinical Investigation*, **56**, 1280–92.

SETO, J. T. & LEPPER, M. H. (1954) The effect of chlortetracycline, oxytetracycline and tetracycline administered intravenously on hepatic fat content. *Antibiotic Chemotherapy*, **4**, 666–72.

SIPE, J. D. (1985) Interleukin 1 as a key factor in the acute phase response. In Gordon, A. H. & Koj, A. (eds), *The Acute Phase Response to Injury and Infection*, pp. 23–35. Amsterdam: Elsevier.

TIMBRELL, J. A. (1983) Drug hepatotoxicity. *British Journal of Clinical Pharmacology*, **15**, 3–14.

(1993) Biotransformation of xenobiotics. In Ballantyne, B., Marrs, T. & Turner, P. (eds), *General and Applied Toxicology*, Vol. 1, pp. 89–120. Basingstoke: Macmillan.

TRYPHONAS, L. & ROZDILSKY, B. (1970) Nuclear jaundice (Kernicterus) in a newborn kitten. *Journal of the American Veterinary Medical Association*, **157**, 1084–7.

TUCKER, R. A. (1982) Drugs and liver disease: a tabular compilation of drugs and the histopathological changes that can occur in the liver. *Drug Intelligence and Clinical Pharmacy*, **16**, 569–80.

WHITBY, L. G., PERCY-ROBB, I. W. & SMITH, A. F. (1984) *Lecture Notes on Clinical Chemistry. Liver Disease*, pp. 169–91. Oxford: Blackwell.

WOODMAN, D. D. (1981) Plasma enzymes in drug toxicity. In Gorrod, J. W. (ed.), *Aspects of Drug Toxicity Testing Methods*, pp. 145–56. London: Taylor & Francis.

(1988) Assessment of hepatic function and damage in animal species. *Journal of Applied Toxicology*, **8**, 249–54.

WOODMAN, D. D. & MAILE, P. A. (1981) Bile acids as an index of cholestasis. *Clinical Chemistry*, **27**, 846–8.

ZIMMERMAN, H. (1974) Hepatic injury caused by therapeutic agents. In Becker, F. F. (ed.), *The Liver*, pp. 225–302. New York: Marcel Dekker.

(1978) *Hepatotoxicity: The Adverse Effects of Drugs and Other Chemicals on the Liver*. New York: Appleton-Crofts.

(1982) Chemical hepatic injury and its detection. In Plaa, G. & Hewitt, W. R. (eds), *Toxicology of the Liver*, pp. 1–45. New York: Raven Press.

7

Assessment of Nephrotoxicity

M. D. STONARD

7.1 Introduction

If the composition of the fluid entering and leaving the kidneys is compared, it is apparent that many of the normal plasma constituents are handled individually, with some undergoing conservation within the tissue and others readily eliminated. For example to illustrate this point, there is a virtual absence of glucose and protein and the appearance of a high concentration of urea in the fluid eliminated from the kidneys. This vital organ not only eliminates urea, the end product of protein catabolism, but also exercises a major influence in regulating the composition of the blood to maintain the internal homeostatic mechanisms. Acid–base, salt and water balances in the body are controlled within the kidneys by the processes of filtration, reabsorption and secretion. Changes in any of these functions will be reflected by a change in the composition of the urine. The kidneys also have an endocrine function; this organ is a major site of synthesis of several hormones including erythropoietin and 1,25-dihydroxycholecalciferol, which influence systemic metabolic functions of the renin-angiotensin and kallekrein-kinin systems (Scicli *et al.*, 1976). These systems are involved in the regulation of blood pressure and vasopressor activity, and of the prostaglandins E_2 and $F_{2\alpha}$, which exert a regulatory effect on sodium homeostasis (Fulgraff and Brandenbusch, 1974).

Many chemical substances have been shown to influence renal function (Commandeur and Vermeulen, 1990). The production of mild and reversible changes is pharmacologically important in the development of diuretics and renin-angiotensin inhibitors; in contrast, irreversible changes may occur following administration of some drugs and chemicals. There are several reasons for the susceptibility of the kidneys to injury. First, renal blood flow is approximately 25% of cardiac output and delivers large quantities of blood to the kidneys, especially the cortex. Second, the kidneys have the ability to concentrate the fluid passing through the tubules such that the concentrations of certain drugs and chemicals may increase 100–1000-fold at certain regions, e.g. proximal tubule. Transport processes for cationic and anionic molecules located in the proximal tubule may be blocked by, or used by, chemicals to gain entry into the renal epithelial cells. Third, the presence in the proximal tubules of mixed-function oxidase enzymes, which have the potential to transform harmless substances into more reactive metabolites, renders this segment of the tubule vulnerable to injury.

7.2 Laboratory Investigations

Several of the current screening methods for nephrotoxicity in animals have evolved from those used in clinical practice. However, it cannot be assumed that a particular analyte or enzyme which has found extensive application in human clinical practice will necessarily have the same diagnostic effectiveness in animal species. Nevertheless, the measurement of certain blood plasma and urinary analytes in combination, can be used to screen for nephrotoxicity (Berndt, 1976; Diezi and Biollaz, 1979; Piperno, 1981; Bovee, 1986; Fent *et al.*, 1988; Stonard, 1990).

When the kidneys are unable to excrete some normal body constituents at the usual rate, the retention of these materials leads to an increase in the plasma concentration, e.g. creatinine and urea. However, urine rather than blood, is the most useful body fluid for the non-invasive investigation of nephrotoxicity. Although urine is a convenient medium to sample, it has to be recognized that the composition of urine can be influenced by organs other than the kidneys. The composition of urine reflects pre-renal, renal and post-renal events which include alterations in hepatic function, e.g. bilirubinuria, changes in carbohydrate metabolism, e.g. glycosuria, and changes in fatty acid metabolism, e.g. ketonuria. Simple tests involve the recording of urine colour and appearance. The normal colour reflects the concentration of urochromes but this can be affected by the presence of both endogenous or exogenous materials, e.g. blood, bilirubin, porphyrins, drugs, dyes. Several of the core urinary measurements, e.g. glucose, blood, pH, etc., can be evaluated qualitatively and/or semi-quantitatively by the use of one of several commercial test strips (or dipsticks). Certain pitfalls may be found with animal species as the test strips are designed for use in human clinical practice. These pitfalls are related primarily to sensitivity and/or specificity, e.g. protein (Evans and Parsons, 1986; Allchin and Evans, 1986b; Allchin *et al.*, 1987). Urine volume and concentration may act as indicators of nephrotoxicity. Many nephrotoxic agents can produce oliguria or polyuria. The latter finding may be a desired pharmacological and reversible response; however, it may indicate an inability by the kidneys to concentrate the tubular fluid. The renal concentrating ability varies from species to species; both the dog and rat can achieve a maximum osmolality approximately twofold greater than in humans (Schmidt-Nielsen and O'Dell, 1961).

In dehydration states, such as that following diarrhoea, it is not uncommon for oliguria to occur until rehydration has taken place. The combined measurement of urinary volume and specific gravity (SG) or osmolality can be used for diagnostic purposes. SG may be measured conveniently with one drop of urine using a refractometer, and osmotic concentration using an osmometer; the presence of glucose and/or protein does not disproportionately affect the measurement. The SG for most laboratory species lies in the range 1.015–1.050. When the body is deprived of water for a period exceeding several hours, the urinary SG and osmolality increase. In renal insufficiency where there is tubular dysfunction, these indices do not increase so greatly. Conversely, when the body is loaded with water, SG and osmolality normally decrease along with an increase in urine flow. In severe tubular disease, the urine tends to remain at a uniform solute concentration, irrespective of whether the urine is formed under conditions of water deficit or excess. The application of a concentration or dilution test may provide a measure of the capacity of the tubules to perform osmotic work. Glycosuria may be detected conveniently by test strips using glucose oxidase methods, although these may be affected by several interferents, including ascorbate. Urinary glucose may reflect enhanced excretion because of elevated blood levels or be due to damage to the proximal tubules, where glucose is reabsorbed.

Changes of hydrogen ion concentration (pH) may reflect changes in tubular function or they may simply reflect dietary protein composition. The pH of urine is a result of its cationic and anionic composition and will be influenced by any substances that perturb the excretion of individual ions, including hydrogen ions. The products of normal metabolism produce an excess of acid substances, which are buffered by a bicarbonate system in the kidney. Urine samples from dog and man are usually acid, containing ammonium ions, due to processes of ionic changes that occur in the distal region of the nephron and also reflecting the nature and composition of the diet being consumed, e.g. protein content.

High concentrations of xenobiotics or metabolites may also result in changes of urinary hydrogen ion concentration. Delays in analysing urine samples may cause hydrogen ion concentration values to be altered by ammonia produced by micro-organisms. Low pH may reflect catabolic states associated with severe toxicity.

The presence of blood in urine may reflect non-specific systemic bleeding, renal or post-renal injury or even an external injury. Some distinction between renal and post-renal injury can be deduced from the combined analysis of haematuria, proteinuria and cast formation. All three components tend to be present in renal injury, whilst haematuria without proteinuria and casts may tend to reflect a post-renal event. A further distinction that can be made about the presence of blood in urine is whether the urine sample is turbid, reflecting haematuria (intact cells), or translucent, reflecting haemoglobinuria (lysed cells); the latter tends to be more common in alkaline or hypotonic urines, or where bleeding occurs higher up the urinary tract. Contamination of urine samples from superficial injuries on paws, etc., should always be considered as a possible explanation for blood in urine.

Microscopic examination of urine sediment for the presence of erythrocytes, leucocytes, renal epithelial cells, bladder cells, spermatozoa, etc., may be used to detect renal tubular damage. The sediment may be examined unstained, although the use of stains helps to distinguish between renal epithelial cells and leucocytes (Prescott and Brodie, 1964; Hardy, 1970). The conditions used for collection, centrifugation and preparation of urine deposits are critical for semi-quantitative analyses.

Renal tubular celluria following a nephrotoxic insult can be differentiated into four phases, (i) initial delay, (ii) peak celluria, (iii) recovery despite continued treatment, (iv) refractory period when re-challenge is ineffective (Prescott and Ansari, 1969).

Timing for urine sampling is therefore critical. The main problem with examination of sediments is quantification. The normal background pattern of celluria differs from one species to another and reference limits have not been established. Methodology is *not* standardized. On the basis of the few published data on urine sediments in species other than man, it must be concluded that celluria is a sensitive but unreliable indicator of acute proximal tubular injury (Davies and Kennedy, 1967; Prescott, 1982). Cell counting is time consuming and special stains are required. The approach appears to be of little benefit for the screening of chronic tubular injury and acute distal tubular damage.

Casts may form in the tubular lumen, and appear to consist of mucoprotein plugs that can trap exfoliated cells and debris. Casts may be hyaline or granular. The major mucoprotein is known to be the Tamm–Horsfall protein, which is located in the distal region of the nephron and is thought to be involved in selective water reabsorption.

Crystals may also be detected microscopically in urine sediments. Their occurrence is largely pH-dependent, e.g. phosphates or urates, and their formation probably corresponds to regions of the nephron where the tubular fluid undergoes significant changes in concentration. Crystals may also reflect high concentrations of an administered substance or metabolite whose solubility characteristics in urine have been exceeded.

Reduced fluid intake or excessive fluid loss by vomitus or diarrhoea will affect urine output. Although measurements of urine cations, e.g. sodium and potassium, and anions may provide further evidence of renal dysfunction, this information may not help to identify the site of the lesion, as reabsorption of electrolytes can occur in various regions of the nephron.

Urinary electrolyte values are highly dependent on dietary intake. Plasma electrolyte measurements may be of some value, but are subject to many extrarenal influences. Urinary magnesium and calcium determination may be important for some nephrotoxic substances, e.g. cisplatin (Magil *et al.*, 1986; see Chapter 12).

Perturbations of acid–base balance will often accompany severe electrolyte changes, but these measurements require controlled conditions, particularly for small laboratory animals. Measurements of renin, aldosterone and atrial natriuretic hormone are generally reserved for xenobiotics where the mode of action suggests that these assays will be useful for the understanding of the pathogenesis of renal changes.

7.3 Proteinuria

Increased urinary protein levels are often the first signs of renal injury. Again test strips offer a simple method for the assessment of proteinuria, but they react mainly with albumin and are therefore not particularly useful in detecting the presence of other urinary proteins which may be found after renal tubular injury. Furthermore, the amounts and composition of the urinary proteins vary between species (Baumann, 1981). The rat shows several qualitative and quantitative differences in protein excretion. The major protein in male rat urine is a low-molecular-weight protein, alpha$_{2u}$-globulin (Roy and Neuhaus, 1967), unlike other species, where albumin is the major urinary protein. Post puberty, male rats excrete more proteins than females and the amount of protein increases with age (Neuhaus and Flory, 1978; Alt *et al.*, 1980). The role of alpha$_{2u}$-globulin in light hydrocarbon nephropathy and hyaline droplet formation has received particular attention in recent years (Stonard *et al.*, 1985; Alden, 1986; Lock *et al.*, 1987). Most of the available quantitative methods for estimating protein in urine have been discussed (Pesce, 1974; see Chapter 14). No one method can be singled out as the technique of choice since all suffer from either lack of sensitivity, endogenous chemical interference or the detection of albumin predominantly. There are several immunoassay techniques available for the measurement of specific proteins in plasma and urine (Woo *et al.*, 1978; Bernard *et al.*, 1981; Viau *et al.*, 1986). However, the selection of methods is governed by the availability of suitably purified proteins and specific antisera for the species of choice.

Primary tubular disorders may be distinguished from glomerular damage by the ratio of high- and low-molecular-weight proteins in the urine. Techniques using various electrophoretic support media and buffers can provide additional information on the composition of urinary proteins (Boesken *et al.*, 1973; Allchin and Evans, 1986a; Stonard *et al.*, 1987).

7.4 Functional Assessment

The kidneys possess the capacity to filter out, via the glomeruli, a whole range of endogenous substances from the plasma, e.g. creatinine, urea, electrolytes, proteins, etc. Since creatinine (end product of creatine catabolism) and urea (end product of protein catabolism) are normally filtered from the plasma and only reabsorbed or secreted by the proximal tubule to a minor extent, both have been used as indices of renal clearance.

The degree of reabsorption or secretion differs from species to species. High values for plasma urea may occur in a variety of conditions other than renal failure, e.g. myocardial infarction, increased dietary nitrogen and gastrointestinal bleeding. Similarly, plasma creatinine concentration may also be affected by diet (Evans, 1987), although this is less well recognized. Plasma creatinine is generally thought to be a better marker than urea of glomerular function. However, plasma creatinine measurements may be subject to analytical interference by endogenous non-creatinine chromogens or some xenobiotics, e.g. some cephalosporins (Evans, 1986; Grotsch and Hajdu, 1987).

An approximation of the glomerular filtration rate (GFR) may be determined by the measurement of endogenous creatinine which involves the collection of timed urine samples. The renal clearance of endogenous creatinine approximates closely to the GFR until renal function is grossly impaired; however, caution is needed as some animal species, e.g. dog, secrete creatinine from the tubule. More reliable measurements of GFR may be obtained by using exogenous inulin with or without radiolabel, [^{51}Cr]-EDTA, or iodothalamate (Chantler *et al.*, 1969; Finco *et al.*, 1981). These determinations require additional infusion and accurately timed blood collection procedures, and may improve the accuracy of GFR determination but reduce the flexibility of study designs. Renal clearance as measured by constant infusion of [^{51}Cr]-EDTA is the same as that for inulin (Chantler *et al.*, 1969). These measurements may be affected by the use of particular anaesthetic agents for blood collection from small laboratory animals. In the dog, inulin clearance is relatively independent of urine flow except in extreme oliguria. Exogenously infused creatinine has also been used to measure renal clearance in the dog (Cronin *et al.*, 1980; Finco *et al.*, 1981).

As glomerular filtration slows or ceases, plasma creatinine and/or urea will increase. The relationship between the functional loss of glomerular filtration and plasma creatinine/urea is not linear. A significant amount of glomerular filtration capacity has to be lost before there is a doubling of plasma creatinine or urea; conversely, plasma urea may remain within the normal range until renal function has been reduced by 50%. The term 'BUN' should be used strictly for the determination of blood urea nitrogen (not plasma or serum urea). Glomerular function, plasma creatinine and urea, to a lesser extent, all show variation with age (Corman *et al.*, 1985; Goldstein, 1990).

The anionic compound, *p*-aminohippuric acid (PAH) can provide an estimate of total renal plasma flow, since it is filtered by the glomeruli and secreted by the renal tubules (Tune *et al.*, 1969). PAH clearance can be used to estimate renal plasma flow, because the combined processes of glomerular filtration and tubular secretion remove PAH entering the kidney by the arterial supply almost completely before it leaves the kidneys. In the dog, approximately 80–90% of PAH in arterial blood is removed before reaching the venous blood supply leaving the kidneys. The proximal convoluted tubule is the primary site for the secretion of PAH. However, PAH clearance does not give an entirely accurate estimate of renal plasma flow, as part of the renal blood flow passes through non-glomerular pathways.

A reduction of PAH clearance may be due to an alteration of renal blood flow by an effect on vasculature or a disruption of the active secretory process. Other organic cations, such as tetraethylammonium (TEA) and phenolsulphthalein (PSP), can also be used as indices of cationic tubular transport systems in the kidney (Plaa and Larson, 1965).

Many substances (both endogenous and exogenous) filtered by the kidneys can be reabsorbed subsequently. Reabsorption can occur passively down a concentration gradient or actively against a concentration gradient; the latter requires energy consumption. Water and chloride ions can be reabsorbed passively whilst reabsorption of amino acids, glucose and sodium ions requires active transport processes.

Amino acids and glucose are completely reabsorbed in the proximal tubule under normal physiological conditions. At low plasma levels, an increased urinary excretion of amino acids or glucose will, in most instances, reflect an impairment of proximal tubular function. At high plasma glucose levels, the tubular reabsorption mechanism may become saturated, with a consequential increase in the urine concentration excreted. The threshold of plasma glucose is normally 10 mmol/L in man and higher in the dog. At plasma glucose levels between 19 and 22 mmol/L, excess glucose is excreted.

Reabsorption of electrolytes, e.g. sodium, can occur in various regions of the nephron. The primary function of the proximal tubules is the iso-osmotic reabsorption of salts and water. Under normal physiological conditions, proximal tubular reabsorption accounts for ca. 50–80% of sodium chloride and water entering the nephron. Such reabsorptive processes have to allow for the principle of electroneutrality, i.e. that net cation reabsorption is equivalent to net anion reabsorption. Thus, the amount of sodium ions reabsorbed is matched by an equivalent reabsorption of chloride ions or other anions.

Table 7.1 Localization of renal tubular enzymes in the rat

Enyzme	EC No.	Localization	Reference
Alanine aminopeptidase (AAP)	EC 3.4.11.1 (cytosol)		
	EC 3.4.11.2 (microsomes)	PST > PCT	Koseki *et al.* (1980)
Alkaline phosphatase (ALP)	EC 3.1.3.1	PCT > PST > distal tubule	Schmidt and Dubach (1971)
β-Galactosidase (GAL)	EC 3.2.1.23	PST > PCT > distal tubule	Lehir *et al.* (1980)
Gamma glutamyl transferase (GGT)	EC 2.3.2.2	PST > PCT > distal tubule	Heinle *et al.* (1977)
Lactate dehydrogenase (LDH)	EC 1.1.1.27	Distal tubule > proximal tubule	Schmidt and Dubach (1971)
β-N-Acetyl glucosaminidase (NAG)	EC 3.2.1.30	PCT > PST = distal tubule	Lehir *et al.* (1980)

PST = proximal straight tubule; PCT = proximal convoluted tubule.

7.5 Enzymuria

Plasma enzyme measurements tend to be insensitive indicators of renal damage because of losses into the lumen of the tubules and hence into the urine. In contrast, urinary enzyme assays are both sensitive and useful for determining the anatomical site of injury (Guder and Ross, 1984). The presence of enzymes in normal urine has been recognized for many years, and reflects cell turnover in the urinary tract. Although enzymes in urine may arise from blood plasma, the detection of abnormally high activities in urine has been interpreted as evidence of renal cellular injury (Wright and Plummer, 1974; Stroo and Hook, 1977). Proof of the renal origin of urinary enzymes has in some instances been demonstrated by a reciprocal reduction in renal tissue enzyme concentration (Ellis and Price, 1975; Cottrell *et al.*, 1976). Several reviews of the diagnostic use of enzymes in urine as indicators of renal damage have been published (Mattenheimer, 1977; Raab, 1980; Price, 1982; Stonard, 1987).

It is clear from these reviews that relatively few enzymes are of diagnostic relevance. Those enzymes that have attracted most attention are those localized in the proximal tubule, a region that is particularly vulnerable to toxic injury (Table 7.1).

The non-linear relationship observed between enzyme activity and urine volume is known in some cases to be due to the presence of endogenous low-molecular-weight inhibitors (Amador *et al.*, 1963) which must be removed by prior dialysis or gel or ultrafiltration (Werner *et al.*, 1969; Werner and Gabrielson, 1977). In some instances, dilution prior to analysis will suffice. This pretreatment, although inconvenient, is essential for most urinary enzymes if erroneous results are to be avoided. The need to remove interfering endogenous substances should be established in preliminary studies. The majority of enzymes show some instability in a hostile medium such as urine, and should be assayed as rapidly as possible after any pretreatment. Also, since the kidney is involved in the elimination of drugs, it may be necessary to perform an *in vitro* incubation of the enzyme with the drug at concentrations expected to be found *in vivo* in order to demonstrate any potential interference. However, this does not eliminate the possibility that enzyme activity may be affected by water-soluble metabolites, whose identity may not have been established.

Several experimental studies have established the role of enzymes in urine, not only as early markers of renal injury (Hofmeister *et al.*, 1986) but also of the site of injury within the nephron. In several acute studies with nephrotoxic agents, it has been possible to demonstrate a direct relationship between urinary enzyme excretion and renal histopathology (Cottrell *et al.*, 1976; Bhargava *et al.*, 1978). In general, those enzymes located on the brush-border region of the proximal renal tubule, e.g. aminopeptidases, alkaline phosphatase and gamma glutamyl transferase (Wright *et al.*, 1972; Jung and Scholz, 1980; Salgo and Szabo, 1982), appear to be earlier indicators of renal damage than other renal tests at the lowest dose levels of nephrotoxins used. However, there are examples where urinary enzymes appear less sensitive than other renal indices (Stroo and Hook, 1977; Kluwe, 1981).

Selectivity of urinary enzymes can be used to distinguish damage to different parts of the nephron. Studies with papillotoxic agents have shown that the production of a voluminous, dilute urine is accompanied by a sustained increase in N-acetyl-β-glucosaminidase excretion, which precedes the excretion of other enzymes indicative of secondary tubular involvement (Ellis *et al.*, 1973; Ellis and Price, 1975; Bach and Hardy, 1985; Stonard *et al.*, 1987).

The use of urinary enzymes to monitor glomerular injury is of limited value. First, few enzymes have been identified as specific markers for glomeruli (Lovett *et al.*, 1982). Second, glomerular damage leads to the loss of enzymes from the blood plasma into the tubular fluid. Thus, it is necessary to discriminate whether an increased excretion of enzyme in urine originates from renal tissue or blood plasma. This may necessitate the development and introduction of methods that discriminate different iso-enzymic forms.

Studies with isolated microdissected nephrons indicate that each segment contains a characteristic pattern of enzymes (Lehir *et al.*, 1980; Guder and Ross, 1984). However, even though several enzymes of the brush-border region have been used diagnostically, differences in the rate of enzyme release from this region of the nephron can be distinguished (Scherberich *et al.*, 1978).

Whilst urinary enzymes have been shown to be of diagnostic value in acute renal tubular injury in animals, their corresponding value in chronic renal injury remains to be established (Plummer *et al.*, 1975). Few experimental models of chronic renal injury have been conducted.

There is evidence of the effects of age, sex, urinary flow and biorhythms on urinary enzyme excretion (Grotsch *et al.*, 1985; Pariat *et al.*, 1990). For some urinary enzymes which exist in iso-enzymic forms, e.g. LDH and N-acetyl-β-glucosaminidase, the measurements can be more informative than that of total enzyme activity (Halman *et al.*, 1984).

References

ALDEN, C. L. (1986) A review of unique male rat hydrocarbon nephropathy. *Toxicological Pathology*, **14**, 109–11.

ALLCHIN, J. P. & EVANS, G. O. (1986a) A simple rapid method for the detection of rat urinary proteins by agarose electrophoresis and nigrosine staining. *Laboratory Animals*, **20**, 202–5.

(1986b) A comparison of three methods for determining the concentration of rat urine. *Comparative Biochemistry Physiology*, **85A**, 771–3.

ALLCHIN, J. P., EVANS, G. O. & PARSONS, C. E. (1987) Pitfalls in the measurement of canine urine concentration. *Veterinary Record*, **120**, 256–7.

ALT, J. M, HACKBARTH, H., DEERBERG, F. & STOLTE, H. (1980) Proteinuria in rats in relationship to age-dependent renal changes. *Laboratory Animals*, **14**, 95–101.

AMADOR, E., ZIMMERMAN, T. S. & WACKER, W. E. C. (1963) Urinary alkaline phosphatase activity. II. An analytical validation of the assay method. *Journal of the American Medical Association*, **185**, 953–7.

BACH, P. H. & HARDY, T. L. (1985) Relevance of animal models to analgesic-associated papillary necrosis in humans. *Kidney International*, **28**, 605–13.

BAUMANN, K. (1981) In Greger, R., Lang, F. & Silbernagl, S. (eds), *Renal Transport of Organic Substances*, pp. 118–33. Berlin: Springer-Verlag.

BERNARD, A. M., VYSKOCIL, A. & LAUWERYS, R. (1981) Determination of β_2-microglobulin in human urine and in serum by latex immunoassay. *Clinical Chemistry*, **27**, 832–7.

BERNDT, W. O. (1976) Renal function tests: what do they mean? A review of renal anatomy, biochemistry and physiology. *Environmental Health Perspectives*, **15**, 55–71.

BHARGAVA, A. S., KHATER, A. R. & GUNZEL, P. (1978) The correlation between lactate dehydrogenase activity in urine and serum and experimental renal damage in the rat. *Toxicology Letters*, **1**, 319–23.

BOESKEN, W. H., KOPF, K. & SCHOLLMEYER, P. (1973) Differentiation of proteinuric diseases by disc electrophoretic molecular weight analytes of urinary proteins. *Clinical Nephrology*, **1**, 311–18.

BOVEE, K. C. (1986) Renal function and laboratory evaluation. *Toxicologic Pathology*, **14**, 26–36.

CHANTLER, C., GARNETT, E. S., PARSONS, V. & VEALL, N. (1969) Glomerular filtration rate measurement in man by the single injection method using ^{51}Cr-EDTA. *Clinical Science*, **37**, 169–80.

COMMANDEUR, J. N. M. & VERMEULEN, N. P. E. (1990) Molecular and biochemical mechanisms of chemically induced nephrotoxicity: a review. *Chemistry Research Toxicology*, **3**, 171–94.

CORMAN, B., PRATZ, J. & POUJEOL, P. (1985) Changes in the anatomy, glomerular filtration rate and solute excretion in aging rat kidney. *American Journal of Physiology*, **248**, 282–7.

COTTRELL, R. C., AGRELO, C. E., GANGOLLI, S. D. & GRASSO, P. (1976) Histochemical and biochemical studies of chemically induced acute kidney damage in the rat. *Food and Cosmetics Toxicology*, **14**, 593–8.

CRONIN, R. E., BULGER, R. E., SOUTHERN, P. & HENRICH, W. L. (1980) Natural history of amonoglycoside nephrotoxicity in the dog. *Journal of Laboratory and Medical Medicine*, **95**, 463–74.

DAVIES, D. J. & KENNEDY, A. (1967) The excretion of renal cells following necrosis of the proximal convoluted tubule. *British Journal of Experimental Pathology*, **48**, 45–50.

DIEZI, J. & BIOLLAZ, J. (1979) Renal function tests in experimental toxicity studies. *Pharmacology and Therapeutics*, **5**, 135–45.

ELLIS, B. G. & PRICE, R. G. (1975) Urinary enzyme excretion during renal papillary necrosis induced in rats with ethyleneimine. *Chemical Biological Interactions*, **11**, 473–82.

ELLIS, B. G., PRICE, R. G. & TOPHAM, J. C. (1973) The effect of tubular damage by mercuric chloride on kidney function and some urinary enzymes in the dog. *Chemical Biological Interactions*, **7**, 101–13.

EVANS, G. O. (1986) The use of an enzymatic kit to measure plasma creatinine in the mouse and three other species. *Comparative Biochemistry and Physiology*, **85B**, 193–5.

(1987) Post-prandial changes in canine plasma creatinine. *Journal of Small Animal Practice*, **28**, 311–15.

EVANS, G. O. & PARSONS, C. E. (1986) Potential errors in the measurement of total protein in male rat urine using test strips. *Laboratory Animals*, **20**, 27–31.

FENT, K., MAYER, E. & ZBINDEN, G. (1988) Nephrotoxicity screening in rats: a validation study. *Archives of Toxicology*, **61**, 349–58.

FINCO, D. R., COULTER, D. B. & BARSANTI, J. A. (1981) Simple, accurate method for clinical estimation of glomerular filtration rate in the dog. *American Journal of Veterinary Research*, **42**, 1874–7.

FULGRAFF, G. & BRANDENBUSCH, G. (1974) Comparison of the effects of prostaglandins A, E_2 and $F_{2\alpha}$ on kidney function in dogs. *Pflugers Archives*, **349**, 9–17.

GOLDSTEIN, R. S. (1990) In Volans, G. N., Sims, J., Sullivan, F. M. & Turner, P. (eds); *Basic Science in Toxicology*, pp. 412–21. London: Taylor & Francis.

GROTSCH, H. & HAJDU, P. (1987) Interference by the new antibiotic cefipirome and other cephalosporins in clinical laboratory test, with specific regard to the Jaffe reaction. *Journal of Clinical Chemistry and Clinical Biochemistry*, **25**, 49–52.

GROTSCH, H., HROPOT, M., KLAUS, E., MALERCZYK, V. & MATTENHEIMER, H. (1985) Enzymuria of the rat: biorhythms and sex differences. *Journal of Clinical Chemistry and Clinical Biochemistry*, **23**, 343–7.

GUDER, W. G. & ROSS, B. D. (1984) Enzyme distribution along the nephron. *Kidney International*, **26**, 101–11.

HALMAN, J., PRICE, R. G. & FOWLER, J. S. L. (1984) Urinary enzymes and isoenzymes of N-acetyl-β-D-glucosaminidase in the assessment of nephrotoxicity. In Goldberg, D. M. & Werner, M. (eds), *Selected Topics in Clinical Enzymology*, pp. 435–44. Berlin: Walder de Gruyter.

HARDY, T. L. (1970) Identification of cells exfoliated from the rat kidney by experimental nephrotoxicity. *Annals of Rheumatic Disease*, **29**, 64–6.

HEINLE, H., WENDEL, A. & SCHMIDT, U. (1977) The activities of the key enzymes of the gamma-glutamyl cycle in micro dissected segments of the rat nephron. *FEBS Letters*, **73**, 220–4.

HOFMEISTER, R., BHARGAVA, A. S. & GUNZEL, P. (1986) Value of enzyme determinations in urine for the diagnosis of nephrotoxicity in rats. *Clinical Chimica Acta*, **160**, 163–7.

JUNG, K. & SCHOLZ, D. (1980) An optimised assay of alanine aminopeptidase activity in urine. *Clinical Chemistry*, **26**, 1251–4.

KLUWE, W. M. (1981) Renal function tests as indicators of kidney injury in subacute toxicity studies. *Toxicology and Applied Pharmacology*, **57**, 414–24.

KOSEKI, C., ENDOU, H., SUDO, J. *et al.* (1980) Evaluation of nephrotoxic site in rat proximal tubule: intrarenal distributions of three enzymes and effects of mercuric chloride and gentamicin on their excretion into urine. *Folia Pharmacology* (Japan), **76**, 59–69.

LEHIR, M., DUBACH, U. C. & GUDER, W. G. (1980) Distribution of acid hydrolases in the nephron of normal and diabetic rats. *International Journal of Biochemistry*, **12**, 41–5.

LOCK, E. A., CHARBONNEAU, M., STRASSER, J., SWENBERG, J. A. & BUS, J. S. (1987) 2,2,4-Trimethylpentane-induced nephropathy II, The reversible binding of a TMP metabolite to a renal protein fraction containing alpha 2μ globulin. *Toxicology and Applied Pharmacology*, **91**, 182–92.

LOVETT, D. H., RYAN, J. L., KASHGARIAN, M. & STERZEL, R. B. (1982) Lysosomal enzymes in glomerular cells of the rat. *American Journal of Pathology*, **107**, 161–6.

MAGIL, A. B., MAVICHAK, V., WONG, N. L. M. *et al.* (1986) Long-term morphological and biochemical observations in Cisplatin induced hypomagnesaemia in rats. *Nephron*, **43**, 223–30.

MATTENHEIMER, H. (1977) Enzymes in renal diseases. *Annals of Clinical Laboratory Science*, **7**, 422–32.

NEUHAUS, O. W. & FLORY, W. (1978) Age-dependent changes in the excretion of urinary proteins by the rat. *Nephron*, **22**, 570–6.

PARIAT, C. L., INGRAND, P., CAMBAR, J., DE LEMOS, E., PIRIOU, A. & COURTOIS, P. T. (1990) Seasonal effects on the daily variations of gentamicin-induced nephrotoxicity. *Archives of Toxicology*, **64**, 205–9.

PESCE, A. J. (1974) Methods used for the analysis of proteins in the urine. *Nephron*, **13**, 93–104.

PIPERNO, E. (1981) In Hook, J. B. (ed.), *Toxicology of the Kidney*, pp. 31–55. New York: Raven Press.

PLAA, G. L. & LARSON, R. E. (1965) Relative nephrotoxic properties of chlorinated methane, ethane and ethylene derivatives in mice. *Toxicology and Applied Pharmacology*, **7**, 37–44.

PLUMMER, D. T., LEATHWOOD, P. D. & BLAKE, M. E. (1975) Urinary enzymes and kidney damage by aspirin and phenacetin. *Chemical Biological Interactions*, **10**, 277–84.

PRESCOTT, L. F. (1982) Assessment of nephrotoxicity. *British Journal of Clinical Pharmacology*, **13**, 303–11.

PRESCOTT, L. F. & ANSARI, S. (1969) The effects of repeated administration of $HgCl_2$ on exfoliation of renal tubular cells and urinary glutamic-oxaloacetic transaminase activity in the rat. *Toxicology and Applied Pharmacology*, **14**, 97–107.

PRESCOTT, L. F. & BRODIE, D. E. (1964) A simple differential stain for urinary sediment. *Lancet*, **288**, 940.

PRICE, R. G. (1982) Urinary enzymes, nephrotoxicity and renal disease. *Toxicology*, **23**, 99–134.

RAAB, W. P. (1980) Nephrotoxicity of drugs, evaluated by renal enzyme excretion studies. *Archives of Toxicology, Supplement* **4**, 194–200.

ROY, A. K. & NEUHAUS, O. W. (1967) Androgenic control of a sex-dependent protein in the rat. *Nature*, **214**, 618–20.

SALGO, L. & SZABO, A. (1982) Gamma-glutamyl transpeptidase activity in human urine. *Clinical Chimica Acta*, **126**, 9–16.

SCHERBERICH, J. E., GAUHL, C. & MONDORF, W. (1978) In Guder, W. C. & Schmidt, U. (eds), *Biochemical Nephrology*, pp. 85–90. Hans Huber: Berne/Stuttgart/Vienna.

SCHMIDT, U. & DUBACH, U. C. (1971) Quantitative histochemie am nephron. *Progress in Histochemistry and Cytochemistry*, **2**, 185.

SCHMIDT-NIELSEN, B. & O'DELL, R. (1961) Structure and concentrating mechanism in the mammalian kidney. *American Journal of Physiology*, **200**, 1119–24.

SCICLI, A. G., CARRETERO, O. A., HAMPTON, A., CORTES, P. & OZA, N. B. (1976) Site of kininogenase secretion in the dog nephron. *American Journal of Physiology*, **230**, 533–6.

STONARD, M. D. (1987) In Bach, P. H. & Lock, E. A. (eds), *Nephrotoxicity in the Experimental and Clinical Situation*, pp. 563–92. Dordrecht/Boston/Lancaster: Martinus Nijhoff.

(1990) Assessment of renal function and damage in animal species. *Journal of Applied Toxicology*, **10**, 267–74.

STONARD, M. D., FOSTER, J. R., PHILLIPS, P. G. N. *et al.* (1985) Hyaline droplet formation in rat kidney induced by 2,2,4-trimethylpentane. In Bach, P. H. and Lock, E. A. (eds), *Renal Heterogeneity and Target Cell Toxicity*, p. 485. Chichester: John Wiley.

STONARD, M. D., GORE, C. W., OLIVER, G. J. A. & SMITH, I. K. (1987) Urinary enzymes and protein patterns as indicators of injury to different regions of the kidney. *Fundamental and Applied Toxicology*, **9**, 339–51.

STROO, W. E. & HOOK, J. B. (1977) Enzymes of renal origin in urine as indicators of nephrotoxicity. *Toxicology and Applied Pharmacology*, **39**, 423–34.

TUNE, B. M., BURG, M. B. & PATLAK, C. S. (1969) Characteristics of PAH transport in proximal renal tubules. *American Journal of Physiology*, **217**, 1057–63.

VIAU, C., BERNARD, A. & LAUWERYS, R. (1986) Determination of rat B_2-microglobulin in urine and in serum, 1. Development of an immunoassay based on latex particles agglutination. *Journal of Applied Toxicology*, **6**, 185–9.

WERNER, M. & GABRIELSON, D. (1977) Ultrafiltration for improved assay of urinary enzymes. *Clinical Chemistry*, **23**, 700–4.

WERNER, M., MARUHN, D. & ATOBA, M. (1969) Use of gel filtration in the assay of urinary enzymes. *Journal of Chromatography*, **40**, 254–63.

WOO, J., FLOYD, M., CANNON, D. C. & KAHAN, B. (1978) Radioimmunoassay for urinary albumin. *Clinical Chemistry*, **24**, 1464–7.

WRIGHT, P. J., LEATHWOOD, P. D. & PLUMMER, D. T. (1972) Enzymes in rat urine: alkaline phosphatase. *Enzymologia*, **42**, 317–27.

WRIGHT, P. J. & PLUMMER, D. T. (1974) The use of urinary enzyme measurements to detect renal damage caused by nephrotoxic compounds. *Biochemical Pharmacology*, **23**, 65–73.

8

Assessment of Endocrine Toxicity

Section 8.1 General Endocrinology

G. O. EVANS

The endocrine system plays an important role in regulating major physiological functions including maturation, reproduction and reactions to external stimuli. Some endocrine organs appear to be more responsive or vulnerable to xenobiotics, and this may result in rapid responses such as those observed for catecholamines and glucocorticoids. However, the majority of effects caused by xenobiotics require longer periods of exposure before the toxic changes are observed, e.g. effects on the pituitary gland or following indirect hormonal changes related to hepatic microsomal enzyme induction. Endocrine changes may be hypo- or hyper-functional due to suppression or following stimulation.

The susceptibility of the endocrine tissues to compound induced lesions has been shown to be ranked in the following decreasing order of frequency: adrenal, testes, thyroid, ovary, pancreas, pituitary and parathyroid (Ribelin, 1984; Woodman, 1988), with an estimated 90% of all toxic effects on the endocrine system involving the adrenals, testes and thyroid glands. Some compounds may affect several endocrine glands, for example by altering both gonadal and thyroid hormones via the pituitary gland. Enhanced hormone secretion appears to promote tumour formation in some cases (IARC, 1979) and it is important to distinguish toxic effects in long-term rat studies where there is a higher spontaneous incidence of usually benign multiple endocrine tumours compared to other species.

Effects of xenobiotics on endocrine organs may become apparent as toxicity or functional impairment. Adverse effects on the endocrine organs are often due to exaggerated responses that can be predicted from knowledge of the pharmacological action(s) of the compound, which has been administered at therapeutic dosages or above. Compounds such as oral contraceptives which are similar to endogenous hormones may mimic a hormone's physiological effects, or structurally similar compounds may block the receptor sites. Endocrine toxins can be broadly categorized as those which exert a direct effect on the hormone-receptor mechanism or those which act via an indirect action.

The indirect effects of xenobiotics on endocrine organs include:

1 inhibition of the enzyme(s) involved in the biosynthesis of the hormone

2 interference in the uptake of the hormone by the target organ
3 interference with the release mechanism for the hormone
4 alteration of capacity of carrier proteins
5 alteration of hormone catabolism, e.g. via hepatic or renal pathways
6 interaction with a secondary messenger system, e.g. cyclic AMP

and several of these are exemplified by cases of thyroid toxicity (see Section 8.2).

Various methods can be used to detect toxic effects on endocrine function, and these include:

1 measurement of endocrine organ mass (absolute and relative to body weight)
2 histological examination of endocrine organ
3 measurement of circulating hormones
4 immunocytochemical examination of endocrine organ
5 surgical removal
6 specific function tests
7 *in vitro* tests
8 indirect biochemical tests

Where endocrine organ toxicity occurs, much of the initial diagnosis still remains with the histopathologist and the recognition of change in organ mass and appearance. Additionally, clinical observations such as fluid intake, lethargy etc. can be useful, and effects on growth or reproductive function can be important clues in endocrine toxicology. The listed methods 3 to 7 are generally selected when further characterization of the mechanistic effect is required following earlier studies, or where the structure of the compound strongly suggests a possible toxic effect. Some hormone measurements may be included in toxicology studies of therapeutic agents which are directed at alterations of endocrine metabolism.

Some effects on the endocrine organs may be detected by other biochemical changes: for example, perturbations of electrolyte balance or carbohydrate metabolism. Nutritional status can have marked effects on hormones which are bound to various plasma protein fractions in the circulation: the relatively weak binding of some hormones to albumin is susceptible to displacement by drugs. Assessment of nutritional status is also important when interpreting results for hormones such as insulin, parathormone and calcitonin.

8.1.1 Hormone Assays

Clearly, the measurements of circulating levels of one or more hormones together with other specific endocrine organ indicators of function are powerful tools in defining the level at which functional impairment occurs. Several of the hormones of interest are shown in Tables 8.1.1 and 8.1.2. Hormones may be measured in blood, plasma (or serum) and urine and immunometric assays have mainly replaced the older hormone bioassays. A full discussion of all

Table 8.1.1 Trophic and releasing hormones, and their abbreviations

Hormone (abbreviation)
Adenohypophysis:
Adrenocorticotrophic hormone (ACTH)
Thyrotrophin or thyroid stimulating hormone (TSH)
Follitrophin or follicle stimulating hormone (FSH)
Luteotrophin or luteinizing hormone (LH)
Somatotrophin or growth hormone (GH)
Prolactin (PRL)
Hypothalamus:
Corticotrophin (corticotrophic) releasing hormone (CRH)
Thyrotrophin (thyrotrophic) releasing hormone (TRH)
Follicle-stimulating releasing hormone (FRH)
Luteinizing-hormone releasing hormone (LRH)
Gonadotrophin releasing hormone (GnRH)
Somatotrophin releasing hormone, somatocrinin (SRH or GRH)
Somatotrophin inhibitory hormone, somatostatin (STH or GIH)
Prolactin inhibitory hormone (PIH; or factor PIH)
Prolactin releasing hormone (PRH)

Table 8.1.2 Classification of some hormones by chemical structure

Protein and peptide hormones:
LH
FSH
Prolactin, inhibin
TRH
CRH
LH-RH
Somatostatin
Prolactin
Amino acid derivatives:
Adrenaline
Noradrenaline
Dopamine
Thyroxine
Tri-iodothyronine
Steroids:
Aldosterone
Cortisol
Corticosterone
Oestradiol
Oestrone
Progesterone
Testosterone

of the different types of immunoassays is not included here. The main methodologies are radioimmunoassay and enzyme-linked immunoassays with their spectrophotometric, fluorescent or luminescent endpoints. A few steroid hormone assays may be determined by other techniques such as high-performance liquid chromatography, spectrophotometric/fluorimetric assays after chemical extraction, e.g. corticosterone (Sargent, 1985), or gas chromatography/mass spectrometry.

Ideally a homologous immunometric assay should be used where the antiserum and standards are available for the particular species. The antibody to a specific hormone of one species may often fail to recognize or bind to the corresponding hormone in another species. As a general rule, antisera for low-molecular-mass hormones such as the steroids, thyroxine, tri-iodothyronine, catecholamines or small peptides will recognize the hormone in several species. More than 30 species share the same sequence for the first 24 of the 40 amino acids residues of ACTH, so most antisera for ACTH will give satisfactory results in these immunoassays. These structural similarities allow several antisera to human hormones to be used with other species.

For some hormones such as insulin and gastrin, the variations in hormonal structure prevent cross-reactivity of certain antisera with some species while allowing others to react. For example, rat and canine plasma but not guinea pig plasma insulin can be measured by human insulin immunoassays.

As the molecular masses of the protein hormones and the differences in amino acid sequences increase, the cross-reactivity of antisera in the different species becomes more problematic, and homologous antisera are required. Several homologous reagents are commercially available, e.g. rat prolactin, thyrotrophin, luteotrophin and follitrophin, canine thyrotrophin. Some specialized centres are able to produce suitable antisera given purified hormones but this is a time-consuming process, and most toxicology laboratories use antisera from external suppliers. Several investigators have reported data for laboratory animals using species-specific antisera provided for research purposes by the National Institute of Arthritis, Metabolism and Digestive Diseases, National Institute of Health, Maryland, USA.

Even when a suitable antiserum has been identified, further technical problems with the immunoassays can occur. The lack of suitable specific standards when using heterologous assays may mean acceptance of a compromise situation, where hormonal changes can be demonstrated but values are expressed in terms of hormone standards obtained from another species. The current lack of specific reagents is an important limiting factor in the development and application of animal endocrinology in toxicology.

Proof of the specificity of an assay is sometimes hindered by the absence of species-specific standards, but some proof can be obtained: preparation of serial dilutions of unknown samples, when added to the antiserum and hormone standards used for the calibration curve, should demonstrate 'dilutional parallelism'. Additional recovery experiments can be performed with known levels of hormones added to the sample matrix. Ideally hormone assays should be evaluated in animal models where effects with test compounds or surgical removal have been described sufficiently well by other investigators. Alternatively, evidence from histopathology examinations can be correlated with hormone measurements.

The calibration curve for human assays may be inappropriate due to the different plasma hormone concentrations in the other species. If the hormone levels are at the lower end of the calibration, the measurements will be less precise than measurements made in the middle of the measuring range. For example, plasma thyroxine (T4) levels are much lower in dog, rat and cat compared with human plasma T4 levels. Eckersall and Williams (1983) using a radioimmunoassay kit for human T4 were unable to detect total T4 in one-third of 'normal' canine samples tested because of these low T4 levels, and they reported that free T4 measurements were more suitable for the detection of canine hypothyroidism. Methods may require modification to allow for these variations in hormone levels. For the hormones that are bound to plasma proteins, any factors affecting protein synthesis and the ratio of free:total hormone levels must be considered.

Plasma hormones exist in a biological matrix which can cause problems in heterologous assays: interactions may occur between the antibody and the plasma proteins, or the variations between the transport binding proteins of different species may also affect the assay. When measuring feline insulin levels with five commercial human insulin assay kits, the results were satisfactory with some kits but problems indicating lack of 'dilutional parallelism' and matrix effects were found with two of the kits (Lutz and Rand, 1993).

8.1.2 Preanalytical Factors Affecting Hormone Assays

Age is an important factor in the assessment of toxic effects on endocrine organs: the developing endocrine system in immature animals appears to be generally more susceptible to toxic compounds than in mature animals, particularly the pituitary–gonadal axis. For some studies, it may be preferable to use male rather than female animals thus avoiding the several metabolic effects of the oestrous cycle.

The effects of stress and wide intra-animal variations often reduce the diagnostic value of hormone assays in laboratory animals. Effects of handling and experimental procedures on prolactin, thyrotrophin, follitrophin, luteotrophin, tri-iodothyronine, and T4 measurements have been described in rats (Wuttke and Meites, 1970; Ajika *et al.*, 1972; Dohler *et al.*, 1977; Gartner *et al.*, 1980), rabbits (Toth and January, 1990), dogs (Garnier *et al.*, 1990), and monkeys (Torii *et al.*, 1993). Examples for other hormones including corticosterone are given in Sections 8.2 and 8.3.

Some hormones show cyclical rhythms and the timing of blood sampling is a critical factor (Kreiger, 1979). Apart from the obvious changes in the levels of sex hormones, particularly in females, periodic variations of other hormones occur in several species. Circadian variations for TSH, T3 and T4 have been described in rats (Jordan *et al.*, 1980) and for testosterone in dogs (Fukuda, 1990). LH, FSH, prolactin and testosterone have been shown to follow annual rhythms in male rhesus monkeys (Beck and Wuttke, 1979). Episodic or pulsatile secretion of plasma prolactin has been observed in rats (Mistry and Voogt, 1989). When cyclical changes do occur, it is difficult to collect samples at the same point in the cycle when dealing with a group of animals, and it may be necessary to take several samples during the cycle. For hormones with circadian rhythms, samples should be taken from a group of animals during a short period within the day.

References

AJIKA, D., KRULICK, L. & MCCANN, S. M. (1972) The effect of pentobarbital (Nembutal) on prolactin release in the rat. *Proceedings of the Society for Experimental Biology and Medicine*, **141**, 203–5.

BECK, W. & WUTTKE, W. (1979) Annual rhythms of luteinizing hormone, follicle-stimulating hormone, prolactin and testosterone. *Journal of Endocrinology*, **83**, 131–9.

DOHLER, K.-D., VON ZUR MUHLEN, A., GARTNER, K. & DOHLER, U. (1977) Effect of various blood sampling techniques on serum levels of pituitary and thyroid hormones in the rat. *Journal of Endocrinology*, **74**, 341–2.

ECKERSALL, P. D. & WILLIAMS, M. E. (1983) Thyroid function tests in dogs using radioimmunoassay kits. *Journal of Small Animal Practice*, **24**, 525–32.

FUKUDA, S. (1990) Circadian rhythm of serum testosterone levels in male beagle dogs – effects of lighting time zone. *Experimental Animals*, **39**, 65–8.

GARNIER, F., BENOIT, E., VIRAT, M., OCHOA, R. & DELATOUR, P. (1990) Adrenal cortical response in clinically normal dogs before and after adaptation to a housing environment. *Laboratory Animals*, **24**, 40–3.

GARTNER, K., BUTTNER, D., DOHLER, K., FRIEDEL, R., LINDENA, J. & TRAUTSCHOLD, I. (1980) Stress response of rats to handling and experimental procedures. *Laboratory Animals*, **14**, 267–74.

IARC (1979) *Monographs on the Evaluation of the Carcinogenic Risk of Chemicals to Humans, Sex Hormones (11)*, 21.

JORDAN, D., ROUSSET, B., FERRIN, F., FOURNIER, M. & ORGIAZZI, J. (1980) Evidence of circadian variations in serum thyrotrophin, 3,5,3′-triiodothyronine, and thyroxine in the rat. *Endocrinology*, **107**, 1245–8.

KREIGER, D. T. (1979) *Endocrine Rhythms*. New York: Raven Press.

LUTZ, T. A. & RAND, J. S. (1993) Comparison of five commercial radioimmunoassay kits for the measurement of feline insulin. *Research in Veterinary Science*, **55**, 64–9.

MISTRY, A. & VOOGT, J. L. (1989) Role of serotonin in nocturnal and diurnal surges of prolactin in the pregnant rat. *Endocrinology*, **125**, 2875–80.

RIBELIN, W. E. (1984) Effects of drugs and chemicals upon the structure of the adrenal gland. *Fundamental and Applied Toxicology*, **4**, 105–19.

SARGENT, R. N. (1985) Determination of corticosterone in rat plasma by HPLC. *Journal of Analytical Toxicology*, **9**, 20–1.

TORII, R., KITAGAWA, N., NIGI, H. & OHSAWA, N. (1993) Effects of repeated restraint stress at 30-minute intervals during 24-hour serum testosterone, LH and glucocorticoids levels in male Japanese monkeys, (*Macaca fuscata*). *Experimental Animals*, **42**, 67–73.

TOTH, L. A. & JANUARY, B. (1990) Physiological stabilisation of rabbits after shipping. *Laboratory Animal Science*, **40**, 384–7.

WOODMAN, D. D. (1988) The use of clinical biochemistry for assessment of endocrine system toxicology. In Keller, P. and Bogin, E. (eds), *The Use of Clinical Biochemistry in Toxicologically Relevant Animal Models and Standardisation and Quality Control in Animal Biochemistry*, pp. 63–77. Basel: Hexagon-Roche.

WUTTKE, W. & MEITES, J. (1970) Effects of ether and pentobarbital on serum prolactin and LH levels in proestrous rats. *Proceedings of the Society for Experimental Biology and Medicine*, **135**, 648–52.

Section 8.2 Thyroid Endocrinology

D. T. DAVIES

In regulatory drug safety evaluation more interest has been shown in the endocrine assessment of the thyroid gland than any other endocrine system. The reason for this interest lies in the observation that the rodent thyroid is particularly sensitive to compound-mediated hormonal disturbance. A variety of compounds from many structural classes and with a variety of pharmacological actions increase thyroid weight and produce follicular cell hypertrophy. Inhibition of thyroid–pituitary homeostasis in long-term studies often leads to the formation of thyroid follicular cell neoplasms. The carcinogenic process proceeds through a number of stages, including follicular cell hypertrophy, hyperplasia, and benign and sometimes malignant neoplasms. In rodents the thyroid gland is the sixth most common site showing carcinogenic activity (Huff *et al.*, 1989) and there is strong evidence to indicate that many thyroid carcinogens act through a 'secondary' mechanism (Capen and Martin, 1989) and should therefore be viewed differently from 'primary' carcinogens for the purpose of risk assessment.

This section reviews the physiology and biochemistry of the thyroid gland, examines the exogenous factors that can influence thyroid–pituitary homeostasis and discusses the scientific approaches that are appropriate to evaluate a novel compound.

8.2.1 Thyroid Hormone Physiology

The thyroid gland plays a crucial role in total body economy and encompasses the synthesis, storage and secretion of the thyroid hormones necessary for growth, development and normal body metabolism. It is generally believed that these actions result from effects of the thyroid hormones on protein synthesis, with the mitochondria as a site of thyroid hormone action. By uncoupling the oxidative phosphorylation process, the thyroid hormones cause increased oxygen consumption and the generation of adenosine triphosphate. Thyroid hormones are also known to stimulate cellular protein and messenger RNA (mRNA) synthesis while stimulating the 'sodium pump' (Na-K-ATPase) of the cell membrane.

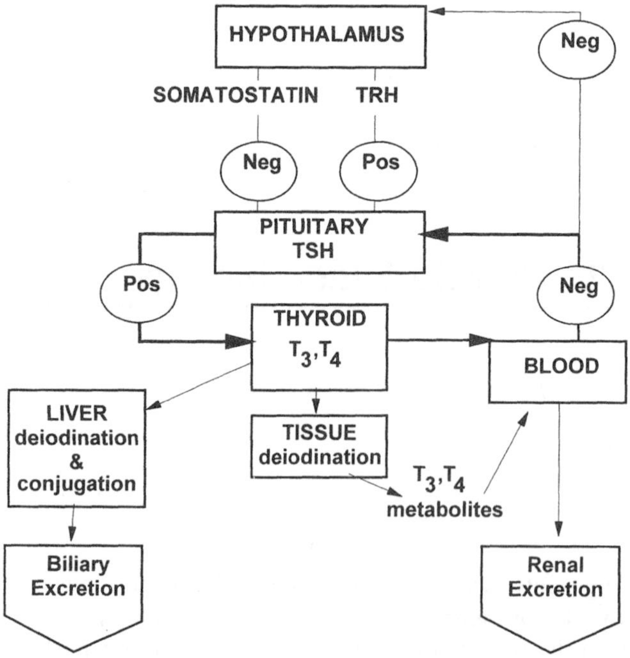

Figure 8.2.1 Control of thyroid gland function

The endocrine control of thyroid gland function is shown schematically in Figure 8.2.1. The synthesis and secretion of the two thyroid hormones, thyroxine (T_4) and tri-iodothyronine (T_3) from the thyroid gland are controlled by thyroid stimulating hormone (TSH) released from the pituitary gland. The pituitary gland is, itself, under the positive control of the thyroid releasing hormone (TRH) and under the negative control of the growth stimulating hormone, somatostatin and the circulating levels of thyroid hormone.

8.2.2 Structure of the Thyroid Gland

The thyroid gland is a bilobed structure which overlays the trachea and is located just below the larynx. The gland is a highly vascularized tissue with a large blood flow. The glandular tissue of the thyroid is composed of follicular cells arranged in closed epithelial-lined spherical structures known as follicles. These possess a central lumen which contains a thick, clear fluid, the colloid. The follicular cells form a single layer of normal cuboidal epithelial cells linked at their apical (inner) surfaces by tight junctions. These cells show microvilli and pseudopodia at the surface in contact with the luminal colloid while their basal surfaces are closely apposed to capillaries. The size and morphology of the follicles vary according to the functional state of the gland. The unstimulated glands show a flattened epithelium and dense colloid. The stimulated gland shows tall, columnar epithelium often with the follicular cells containing colloid droplets, together with a watery colloid in the follicular lumen. The colloid space is reduced and the highly stimulated gland consists largely of highly active tall columnar epithelial cells forming small follicles with small colloid spaces. Interspersed among the follicles are the thyroid C (parafollicular) cells. A more detailed description of thyroid structure is given in Greaves (1990).

8.2.3 Calcitonin

The functions of the C cells are the synthesis, storage and release of calcitonin in response to an appropriate physiological stimulus. There are numerous calcitonin-containing granules in the cytoplasm of C cells, and these accumulate in response to hypocalcaemia and degranulate in hypercalcaemia. Unlike the follicular cells, which are responsive to a pituitary hormone, calcitonin is secreted primarily in response to altered circulating calcium levels. Calcium homeostasis in extracellular fluids is maintained by calcitonin functioning as a hormone which prevents the development of hypercalcaemia during the rapid postprandial absorption of calcium via the gastrointestinal tract (see also Chapter 9). A more detailed description of the role and action of calcitonin in animals is presented by Simesen (1980).

8.2.4 Thyroid Hormone Synthesis

It is important to briefly describe the physiology and biochemistry of the thyroid–pituitary hormonal system, before discussing how xenobiotics alter thyroid endocrine homeostasis.

The first stage in the synthesis of the thyroid hormones is the uptake of iodine from the blood by the thyroid gland (Figure 8.2.2). Uptake is active in nature (requires energy) and is achieved by the so-called 'iodine pump'. Under normal conditions the thyroid may concentrate iodide up to about 50-fold its concentration in blood, and this ratio may be considerably higher when the thyroid is active. Once trapped in the cell the next stage involves the oxidation of iodide (I^-) to iodine (I_2) by a peroxidase in the presence of hydrogen peroxide. The amino acid tyrosine, transported from the bloodstream, is also needed since the active iodine species iodinates the tyrosyl residues of thyroglobulin to form mono-iodotyrosine (MIT) and di-iodotyrosine (DIT). The iodinated tyrosines are next coupled to form either L-3,5,3′-tri-iodotyrosine (T_3) or L-3,5,3′,5′-tetra-iodotyrosine (T_4, thyroxine), the principal thyroid hormones. It is thought that these reactions are catalysed by the same peroxidase effecting the iodination reaction and occur in the follicular lumen. These thyroid hormones, bound to thyroglobulin, are stored in the follicular lumen as colloid. The release of T_4 and T_3 from thyroglobin is effected by endocytosis of colloid droplets into the follicular epithelial cells and subsequent action of lysosomal proteases. The free hormones are subsequently released into the circulation while MIT and DIT are catabolized and iodine is recirculated. About 90% of the hormone released into the circulation is T_4, which is usually considered to be a prohormone for T_3. Tri-iodotyrosine (T_3) is about fourfold more potent than thyroxine (T_4), and about 33% of the T_4 secreted undergoes 5′-deiodination to the active T_3 in the peripheral tissues. Inner ring deiodination of T_4 can also occur with the formation of the biologically inactive reverse T_3 (rT_3).

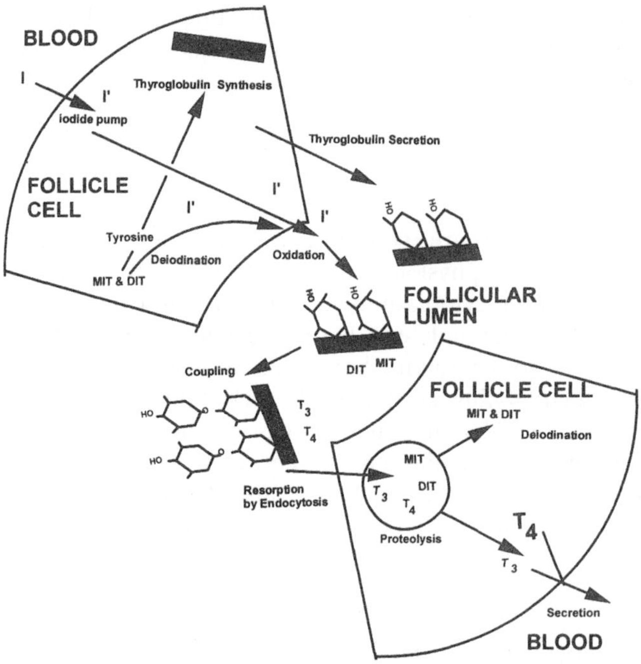

Figure 8.2.2 Schematic representation of thyroid hormone biosynthesis and secretion

In all species the vast majority of the circulating thyroid hormones are protein bound but the proportion of high-affinity/low-capacity and low-affinity/high-capacity binding differs between species (Woodman, 1988). In man, the majority of circulating thyroid hormones are bound to the high-affinity thyroglobulin (TBG) molecules but in rodents this molecule is absent and low-affinity/high-capacity binding is provided mainly by albumin and prealbumin. Species differences in the levels of total and free T_3 and T_4 have been found in the serum of cattle, goats, guinea pigs, horses, pigs, rats and sheep (Refetoff *et al.*, 1970; Anderson *et al.*, 1988). Free and bound T_3 and T_4 are in dynamic equilibrium in the circulation and because of the differences in affinity for the binding proteins there is more T_3 in the unbound form (ca. 4% of total unbound) than T_4 (ca. 0.04% of total unbound). It is the unbound form of each hormone which is metabolically active.

The thyroid stimulating hormone (TSH), released from the pituitary gland, stimulates most aspects of thyroxine synthesis and secretion. TSH increases iodide uptake, iodination of thyroglobulin, endocytosis of thyroglobulin and, finally, proteolysis of the colloid with release of thyroid hormones. The rate of release of TSH from the pituitary is finely controlled by the amount of thyrotropin releasing hormone (TRH) secreted by the hypothalamus and by the circulating levels of T_4 and T_3. As shown in Figure 8.2.3, if there is a decrease in the circulating levels of thyroid hormones, TSH is secreted and thyroid function is increased; if, on the other hand, circulating thyroid hormone levels increase, TRH secretion is suppressed and eventually the thyroid gland becomes inactive and regresses. This dynamic feedback mechanism functions to maintain a steady secretion and stable circulating levels of thyroid hormones.

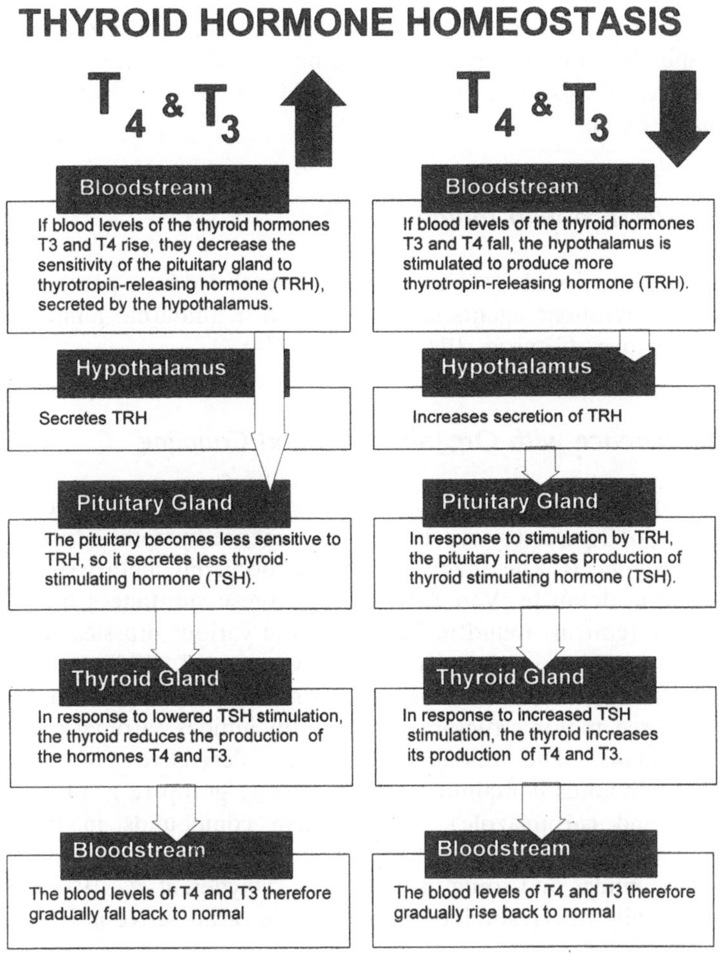

THYROID HORMONE HOMEOSTASIS
T4 & T3
T4 & T3
Bloodstream
If blood levels of the thyroid hormones T3 and T4 rise, they decrease the sensitivity of the pituitary gland to thyrotropin-releasing hormone (TRH), secreted by the hypothalamus.
Hypothalamus
Secretes TRH
Pituitary Gland
The pituitary becomes less sensitive to TRH, so it secretes less thyroid stimulating hormone (TSH).
Thyroid Gland
In response to lowered TSH stimulation, the thyroid reduces the production of the hormones T4 and T3.
Bloodstream
The blood levels of T4 and T3 therefore gradually fall back to normal
Bloodstream
If blood levels of the thyroid hormones T3 and T4 fall, the hypothalamus is stimulated to produce more thyrotropin-releasing hormone (TRH).
Hypothalamus
Increases secretion of TRH
Pituitary Gland
In response to stimulation by TRH, the pituitary increases production of thyroid stimulating hormone (TSH).
Thyroid Gland
In response to increased TSH stimulation, the thyroid increases its production of T4 and T3.
Bloodstream
The blood levels of T4 and T3 therefore gradually rise back to normal

Figure 8.2.3 Thyroid gland homeostasis

8.2.5 Perturbation of Thyroidal Function

The process of thyroid hormone synthesis is complex and highly susceptible to disruption by xenobiotics. The classification of reported xenobiotic effects on thyroid function has recently been reviewed by Atterwill and colleagues (1992) and for further details the reader is referred to Cavalieri and Pitt-Rivers (1981) and Capen and Martin (1989). A brief summary of the possible perturbation sites with a few key examples follows.

8.2.5.1 Interference with Iodide Trapping

The iodide-concentrating mechanism, which resides within the follicular epithelial membrane, can concentrate iodide to levels approximately 30 times higher than those of plasma. Thyroid stimulating hormone enhances the transport mechanism whereas inorganic ions such as thiocyanate, perchlorate, dithiocarbamate fungicide (Nabam), and dichlorodiphenyl trichloroethane (DDT) inhibit iodide trapping (Goldman *et al.*, 1970; Netter, 1974; Kuzan and Prahlad, 1975; Bowman and Rand, 1980; Wenzel, 1981).

8.2.5.2 ***Interference with Iodide Oxidation***

The oxidation of the iodide (I^-) to iodine (I_2) catalysed by a peroxidase reaction is inhibited by thyrotoxic agents such as thiouracil, thiourea, aminotriazole and imidazole derivatives (Taurog, 1970; Netter, 1974).

8.2.5.3 ***Interference with Organification and Coupling***

The iodination of tyrosyl residues to form MIT and DIT and the subsequent formation of T_3 or T_4 are very sensitive to interference by both naturally occurring and synthetic chemicals. The effects of naturally occurring dietary substances have been reviewed in detail by Van Etten (1969): these substances include vinyl-2-thiooxazolidone (goitrin) found in rape seed and various brassicas (Griesbach *et al.*, 1945), and arachidoside in peanuts (Srinivasan *et al.*, 1957).

Synthetic chemicals that interfere with organification and coupling can be divided into three major structural groups: thioamides, aromatic amines, and polyhydric phenols. The thioamides include derivatives of thiourea and heterocyclic compounds containing the thiourelyne group (e.g. propylthiouracil, methimazole, and carbimazole). Other active compounds in this class are derivatives of imidazole, oxazole and thiazole. Examples of the aromatic amines are the sulphonamides and some oxydianilines (Hayden *et al.*, 1978; Haynes and Murad, 1985) while hexyresorcinol is an example of an active polyhydric phenol (Paynter *et al.*, 1986).

8.2.5.4 ***Effect on Hepatic Thyroid Hormone Metabolism***

Equally profound disturbances in thyroid function and morphology can occur with chemicals that have no direct effect on the thyroid gland. Of particular interest are those compounds that induce a variety of hepatic and/or extrahepatic enzymes responsible for the metabolism of several endogenous and exogenous compounds. In the rat, approximately half of the T_4 eliminated from the body occurs via the bile. The major pathway of elimination involves conjugation of the phenolic hydroxyl group of T_4 with glucuronic acid and biliary excretion of the resulting glucuronide (Galton, 1968; Bastomsky, 1973); sulphate conjugates may also be produced and excreted. Several compounds can induce the conjugating enzyme uridine disphosphate glucuronyl-transferase (UDP-GT), and can increase clearance of T_4 from the circulation with increased biliary excretion of the T_4 glucuronide-conjugate.

Other enzyme-inducing compounds can be classified according to their effect on cytochrome P450 and mono-oxygenase activity (Mannering, 1971). One class, typified by phenobarbitone, produces a significant increase in liver weight and a significant proliferation of hepatic endoplasmic reticulum. Induction is associated with increases of hepatic cytochrome P450 and a large number of oxidative associated with mono-oxygenases. The chlorinated hydrocarbon insecticides (DDT, chlordane and aldrin) exhibit induction characteristics similar to phenobarbitone.

The effect of phenobarbitone-like inducers on thyroid function are known to be complex. Animals treated with phenobarbitone show increased hepatocellular binding of T_4 combined with enhanced biliary excretion of the hormone (Oppenheimer *et al.*, 1968, 1971). In rats, these changes result from an increased rate of turnover of T_4 that is compensated by the release of TSH and enhanced secretion of more thyroid hormone.

The second class of inducers, typified by the polycyclic hydrocarbon, 3-methyl cholanthrene (3MC), do not cause an increase in liver size and increase the formation of the cytochrome P448 (Ryan *et al.*, 1978). Inducers of the '3MC type' include a number of polycyclic aromatic hydrocarbons, naphthoflavone, and several halogenated dibenzo-*p*-dioxins (2,3,7,8-tetrachlorodibenzo-*p*-dioxin, TCDD). With the '3MC-type' inducers, the major mechanism appears to be the induction of T_4-UDP-glucuronyl-transferase, which is the rate-limiting step in the biliary excretion of T_4 (Bastomsky, 1973); this enzyme induction results in a lowering of circulating plasma T_4 levels (Potter *et al.*, 1983).

Finally, because of their widespread environmental use, mention must be made of the thyroid effects of the PCBs and PCCs (the polychlorinated and polybrominated biphenyls). These polyhalogenated biphenyls show the characteristics of both phenobarbitone and the 3MC-type inducers (Alvares *et al.*, 1973), with a greater than threefold decrease in plasma T_4 levels noted in PCB-treated rats (Collins *et al.*, 1977).

8.2.5.5 Effect on 5′-monodeiodinase

Certain thionamides, in addition to their known inhibitory effect on iodination and coupling of tyrosine moieties to form T_4 and T_3, also have the ability to inhibit the peripheral conversion of T_4 to T_3. With these compounds the pattern of circulating hormone changes that can be seen is reduced circulating T_3, increased TSH and reverse T_3 (rT_3) with normal or elevated T_4. (As mentioned previously, T_4 is considered to be a prohormone and is converted to the more active T_3 in peripheral tissues, which include the pituitary.)

Hypothyroidism has been observed in man following the administration of amiodarone, where serum TSH levels were doubled (Borowski *et al.*, 1985). Similar changes have been observed in rats dosed with the colour additive Red No. 3 (Erythrosine) where lowered T_3 values, elevated T_4 values, increased rT_3 and increased TSH levels suggest a peripheral metabolizing enzyme inhibitory effect for this compound (Capen and Martin, 1989).

8.2.5.6 Effects on Plasma Protein Binding

In man, circulating thyroxine is bound to a specific, high-affinity protein called thyroxine binding globulin (TBG). Because of this high affinity thyroxine has a circulating half-life in man of between 5 and 9 days. In rodents, this protein is absent and 75% of thyroxine is bound to albumin. This has a low affinity for thyroxine and, consequently, the half-life of T_4 in the rat is about 12 to 16 hours. In addition, the ratio of tissue volume/plasma flow rate of the liver in the rat is 2.5 times higher than in man (Latropoulos, 1993/94). These observations indicate a much higher turnover of the thyroid hormones in rats and the serum TSH is 25 or more times higher in the rat compared with man (Dohler *et al.*, 1979). This indicates a much higher activity in the rat thyroid gland compared with primates, a conclusion supported by the histomorphology of the rat thyroid gland which often appears to be hypertrophic and hyperplastic, even in healthy untreated rats.

This is supported by the observation that rats require a 10-fold higher T_4 production rate per kilogram body weight than do humans to maintain physiological plasma levels (Dohler *et al.*, 1979). Several compounds have an indirect effect on thyroid hormone binding. These compounds include phenytoin which reduces the thyroid hormone binding to plasma protein in both rat and man, and salicylates which reduce thyroxine binding to TBG and the low-affinity thyroid binding pre-albumin (TBPA) (Cavalieri and Pitt-Rivers, 1981; Wenzel, 1981).

8.2.6 Consequences of Prolonged Elevated TSH Secretion

Increased TSH secretion due to a reduction in circulating thyroid hormones, caused either by direct actions of xenobiotics or as a consequence of increased hormone clearance, are significantly more common in animals than as a result of a primary effect on the pituitary gland. Regardless of this secondary mechanism, the response to hypothyroidism is similar and the biological consequences of prolonged elevated TSH secretion have been extensively described (Wynford-Thomas *et al.*, 1982a, 1982b, 1982c; Stringer *et al.*, 1985; Smith *et al.*, 1986; Williams *et al.*, 1988).

As a consequence of increased TSH secretion the thyroid gland undergoes rapid growth, the colloid in the follicular lumen declines, and this leads to follicular hypertrophy. In the rat, this growth phase can last for up to 3 months, at which time the gland becomes desensitized to further TSH stimulation and growth stops; how this regulation is achieved is not known. It is recognized that TSH acts in concert with various growth factors, such as insulin-like growth factor 1 (IGF_1), to regulate normal cell proliferation (Smith *et al.*, 1986; Williams *et al.*, 1988). Other stimulatory growth factors may be involved, e.g. IGF_2, epidermal growth factor (EGF), whereas inhibitory factors, such as transforming growth factor β (TGFβ), have been shown to inhibit the stimulatory effect of TSH (Grubeck-Loewenstein *et al.*, 1989).

When the thyroid gland becomes desensitized to the stimulatory effect of TSH, it enters a 'plateau growth phase' which can persist for up to a year. During this time, although there is no outward appearance of change, it is believed that spontaneous mutations occur within the gland that give rise to one or more clones of cells, some of which, through a further mutation, can become TSH sensitive and acquire an additional growth advantage. The finding of thyroid neoplasms in about 1% of some laboratory animals (Haseman *et al.*, 1984) supports the idea that 'spontaneous mutations' may occur in control animals and lead to the

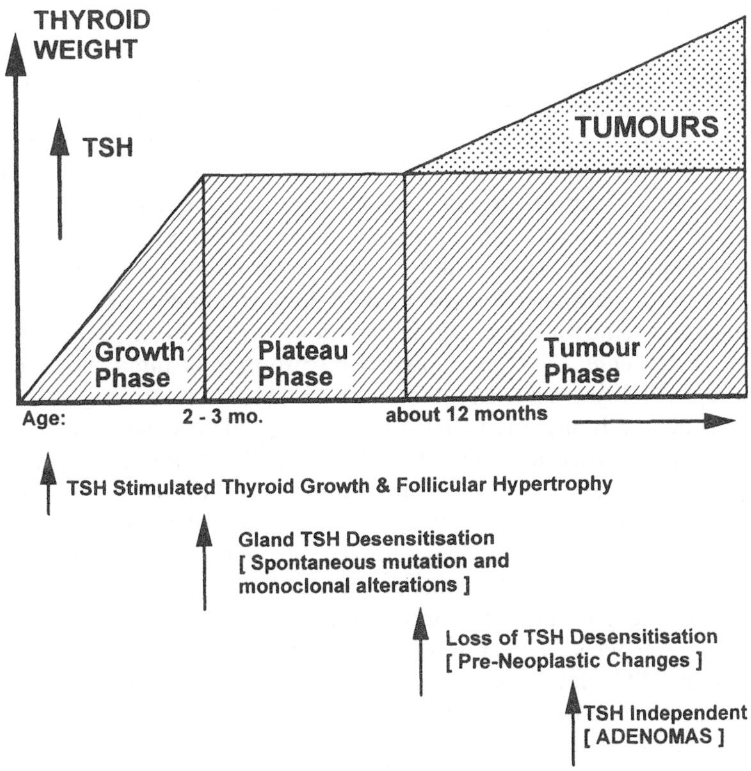

Figure 8.2.4 Thyroid follicular adenoma progression

development of thyroid tumours. Once activated, these cell clones undergo a stepwise progression from hypertrophy and hyperplasia, nodular hyperplasia, and then to neoplasia (see Figure 8.2.4). If the administration of the goitrogen is discontinued and TSH secretion is allowed to return to normal levels, the progression towards tumour formation can be reversed and the thyroid gland will return to normal (Thomas *et al.*, 1991). This reversal can be achieved any time up to about one year from the start of goitrogen administration in the rat; during the second year, the growth of the nodules becomes TSH independent and development of adenomas and carcinomas ensues, irrespective of treatment.

Unlike humans, rodents appear to be highly responsive to these TSH-induced proliferative changes and this results in an increase in the occurrence of thyroid adenomas, as well as follicular hyperplasia and follicular cysts (Capen, 1983). Excessive secretion of endogenous TSH alone, in the absence of any chemical treatment, will produce a high incidence of thyroid tumours in rats fed iodine-deficient diets (Bielschowsky, 1953; Axelrad and Lebond, 1955) or in mice transplanted with a TSH secreting pituitary tumour (Furth, 1954). In contrast, endemic goitre has affected tens of millions of people from iodine-deficient areas but a clear aetiologic role in thyroid gland neoplasia has not been established (Pendergrast *et al.*, 1961; Doniach, 1970).

8.2.7 Assessment of a Potential Thyroid Carcinogen

It is generally accepted that carcinogenesis is a multistage process that usually includes an initiation step as well as a promotional phase (OSTP, 1985). Treatment with the first agent, a genotoxin or mutagen, confers a permanent change on the cells such that prolonged exposure to a second agent, the promoter or goitrogen, results in neoplasms. Although goitrogenic stimuli that increase TSH levels (e.g. iron deficiency, treatment with propylthiouracil, PTU) are known to induce thyroid hyperplasia and neoplasia alone, many experiments have demonstrated an enhancement of the neoplastic response when these treatments are combined with other exposures to known direct-acting carcinogens (e.g. N-bis[2-hydroxypropyl] nitrosamine, Hiasa *et al.*, 1982; N-methyl-N-nitrosourea, Oshima and Ward, 1984; dihydroxy-propyl-nitrosamine, McClain *et al.*, 1988).

While it is generally accepted that a hormonal feedback mechanism involving increased output of thyroid stimulating hormone from the pituitary gland in response to low hormone levels may be implicated in the formation of thyroid follicular cell tumours (Woo *et al.*, 1985; Paynter *et al.*, 1986), it is also important to evaluate the mutagenic/genotoxic potential of all agents causing thyroid tumours in order to discount any direct genotoxic effect.

Hill and his colleagues (1989), working under the auspices of the Environmental Protection Agency's (EPA) Risk Assessment Forum, have examined the hypothesis that TSH plays a significant role in thyroid carcinogenesis and that chemicals that produce thyroid tumours in experimental animals would not show genotoxic potential in any predictable way. Following detailed examination of three classes of compounds: thionamides, aromatic amines and halogenated hydrocarbons, they conclude that 'within the limits of the present review, there does not seem to be a consistent relationship across chemical classes that produce thyroid tumours and their ability to produce genotoxic effects'. In man, no relationship between hypothyroidism and thyroid cancer has been described (McTiernan *et al.*, 1984; Ron *et al.*, 1987) and there is considerably less support for a role of TSH in human thyroid carcinogenesis, compared with that for rodents.

Although it is never easy to prove that carcinogenesis proceeds via specific discrete steps, in certain cases there are enough experimental data to imply that a certain sequence of events occurs, and this information can be used as the basis for an approach to estimate human cancer risk. Studies over the last 30 years in many laboratories, using a number of different study protocols, indicate that this is the case for the induction of certain follicular cell neoplasms of the thyroid gland. Clearly, if a compound produces tumours at sites other than the thyroid (or pituitary); has a genotoxic potential or does not seem to be acting via thyroid–pituitary hormone inhibition, then conventional principles laid down by various regulatory authorities for Carcinogen Risk Assessment must be applied. Where it is reasonable to presume that the neoplasms are due to thyroid–pituitary hormonal imbalance and where other carcinogenic mechanisms can be discounted (e.g. genotoxicity), then the compound can be considered as posing little or no risk to man provided it can be demonstrated that:

1 The compound has goitrogenic activity *in vivo* (i.e. thyroid follicular hypertrophy and hyperplasia).

2 There are clinical chemistry indications of changes in thyroid and pituitary function (i.e. reduced thyroid hormone and increased TSH plasma concentration).
3 There is specific evidence that the agent either reduces thyroid hormone synthesis (e.g. inhibits iodine uptake) or increases thyroid hormone clearance (e.g. enhanced biliary excretion).
4 A progression of lesions in studies of various duration, showing cellular hypertrophy and hyperplasia, nodular hyperplasia, and neoplasia (benign and possible malignant tumours) can be demonstrated.

As for other carcinogens, it is important to quantify the risk to man and establish the threshold dose levels for the various toxicological end-points (e.g. thyroid weight increase, follicular cell hypertrophy, plasma thyroid hormone changes; see Paynter *et al.*, 1988).

8.2.8 Laboratory Assessment of Thyroidal Status

Organ weight changes and histopathological assessments can provide good evidence that the thyroid status has been altered but never indicate the mechanism by which the xenobiotic has produced the change. If during any stage of a novel compound's safety assessment it is found to cause thyroid hypertrophy in rats, then it is prudent to undertake an assessment of thyroid–pituitary hormonal status.

Several approaches should be considered and, in addition to measuring plasma hormone concentrations, it is important to identify the mechanism by which the compound has produced its effect.

8.2.8.1 TSH and Thyroid Hormone Evaluation

Several commercial immunoassay kits are available for the assessment of human T_3 and T_4, and since these hormones have a common structure across species these assays can be reliably used to measure thyroid hormones in the rat. Other methods for assessing rat thyroid hormones have been described by Kaneko (1980) but these are probably less sensitive than the radioimmunoassay methods.

Thyroid stimulating hormone (TSH) is a glucoprotein whose structure is slightly altered in different species. Since all sensitive assays involve the use of peptide specific antibodies, a rat-specific TSH assay is required to reliably measure this hormone in rodents. Antibodies and purified rat-TSH are available for scientific research from the National Hormone Programme of NIDDK (Baltimore, Maryland) while commercially prepared rat-TSH kits are available from Amersham International and Diagnostic Products Corporation (DPC) Ltd. The DPC rat-TSH kit is a non-isotopic, automated assay utilizing an immunofluorescence marker. Experience, to date, indicates that the DPC assays are reliable and provide sound results for the evaluation of thyroid hormones in rodents.

Goitrogenic xenobiotics that have a direct inhibitory effect on thyroid hormone synthesis (e.g. propylthiouracil, PTU) produce marked plasma hormone changes in rodents. These plasma thyroid hormones and TSH changes are very easily demonstrable with a greater-than-80% reduction in rat plasma thyroxine and a sevenfold increase in TSH being observed after five daily doses of PTU (Davies,

1993). Species differences have been noted in response to direct-acting goitrogens and rodents appear to be particularly susceptible to these compounds. Following the daily administration of ethylene thiourea to rats and mice, thyroxine but not T_3 was suppressed throughout the 90-day duration of the study. Plasma TSH values, in contrast, were significantly elevated at 7, 28 and 90 days in the mouse but, in the rat, following a peak at day 7, had returned to control values by day 90, despite the continued suppression of both total and free T_4 (Elcombe, personal communication). However, in the beagle dog, daily administration of PTU at the compound's maximally tolerated oral dose for 28 days, failed to modify plasma T_3 and T_4 values until at least day 14. A slight increase in plasma TSH values was apparent after a further 3 days and the maximum value of TSH achieved was approximately double that observed pre-study.

For compounds that have a secondary thyroid effect, as a consequence of increased thyroxine clearance, changes in hormone levels are more modest. Seven days after the administration of the inotropic drug, OPC 8212, serum T_4 was reduced by about 20% but TSH values were more than doubled (Lueprasitsakul *et al.*, 1991). Similarly, McClain (1989) was able to demonstrate that phenobarbitone, by its liver microsomal enzyme-inducing effect, also modified thyroid–pituitary homeostasis. After 2 weeks of dosing, a 30% fall in serum T_4 was seen in male rats and this was accompanied by a doubling of serum TSH concentration. Over the following 10 weeks, serum TSH and T_4 concentrations returned to near pre-study values although thyroid weights remained elevated.

When examining the effect of the UDP-glucuronyltransferase inducers, phenobarbital (PB), 3-methylcholanthrene (3MC), pregnenolone-16α-carbonitrile (PCN) and a mixture of polychlorinated biphenols (PCBs), on thyroid function, Barter and Klaassen (1994) demonstrated a correlation between UCP-GT activity towards T_4 and the reduction in serum thyroxine. The reductions in thyroid hormone levels led to an increase in serum TSH and increased thyroid weight; however, no clear correlation between lower serum thyroxine values and increased serum TSH values was apparent. While PB, 3MC and PCN produced a fall in T_4 of between 30 and 40%, the polychlorinated biphenols produced an 80–90% reduction. With respect to the TSH increase, PB, 3MC and the PCBs increase TSH by 40–50% but PCN, which had a modest effect on T_4, produced a 210% increase in TSH. Similar results have been described by Takaoka *et al.* (1994) who demonstrated that while the 3-nitro-triazole produced a 41% reduction in T_4 and a threefold increase in TSH, the 3-mercapto-triazole produced a 50% reduction in T_4 but a ninefold increase in TSH. Neither compound had any effect on the thyroid–pituitary axis other than the inhibition of thyroid hormone synthesis.

The above examples serve to illustrate the degree of variability that can be expected when measuring plasma or serum thyroid hormones and TSH in the rat. Personal experience of evaluating thyroid endocrine changes with compounds known to cause thyroid follicular hypertrophy have, on occasions, been somewhat inconclusive (Davies, 1993). To minimize the difficulties associated with large inter-animal variability every effort must be made to reduce the impact of environmental and technical variables. In order to improve the diagnostic accuracy and consistency of thyroid hormone assessment, a number of measures have been proposed (Davies, 1993):

1 Use at least 20 rats per sex per group.

2 Collect blood samples at necropsy rather than multiple interim blood collections.
3 Confine the time for collecting the blood to a 1-hour period thus avoiding problems associated with circadian rhythms.
4 Samples should be collected at regular intervals after the start of dosing in order to observe plasma/serum changes prior to onset of homeostatic regulation.
5 Store the samples at −20 or −70°C and analyse all samples within a single assay run at the end of the study.

8.2.8.2 *Plasma Hormone Clearance Studies*

Compound-induced changes in hepatic microsomal enzyme activity, especially that of thyroxine glucuronyltransferase activity, and changes in bile flow can have significant effects on thyroid hormone metabolism. Direct measurement of plasma hormone clearance and T_4 half-life can be determined in compound-dosed rats by measuring the plasma values of ^{125}I-T_4 following the intravenous bolus injection of radioactive thyroxine. Either intact (Davies, 1993) or thyroidectomized (McClain *et al.*, 1989) rats can be used and while determining the total radioactivity of plasma is acceptable, a more accurate approach is to count specific ^{125}I-bound-T_4 by binding the hormone to T_4 antibodies attached to the surface of polypropylene tubes. Following the administration of an anti-inflammatory agent known to be a liver enzyme inducer, Davies (1993), using the latter approach, demonstrated a reduction in plasma T_4 half-life and an increase in the plasma clearance of thyroxine from 2.86 ml min^{-1} kg^{-1} to 3.37 ml min^{-1} kg^{-1} ($p<0.001$). In this study no clear change in plasma thyroid hormones or TSH was demonstrated.

8.2.8.3 *Radioactive Iodide Incorporation Studies*

Brown *et al.* (1987) have examined the effect of the histamine antagonist, Lupitidine, on thyroid function. Lupitidine caused a dose-dependent increase in thyroidal iodine incorporation that was apparent after a single dose of the compound; this effect was reversible after seven days' dosing. The increased incorporation of ^{125}I into the thyroid gland was apparently dependent on TSH since both hypophysectomy and pretreatment with thyroxine markedly reduced thyroidal ^{125}I uptake.

8.2.8.4 In Vitro *Thyroid Hormone Studies*

In vitro toxicological studies with primary cultures of porcine thyrocytes have provided evidence for good correlation with the *in vivo* potency of various xenobiotics (Brown, 1987; Reader, 1987; Atterwill and Fowler, 1990). This is a useful technique for examining compounds that are suspected of interference with organification, and these *in vitro* effects can be measured either by iodide incorporation or the perchlorate discharge test (Atterwill *et al.*, 1987).

The direct effect of xenobiotics on either rat pituitary thyrotrophs or thyrocytes can also be evaluated by measuring the amount of TSH or thyroxine in the *in vitro* culture after an appropriate period of incubation.

8.2.9 Conclusions

There is now a wealth of literature devoted to understanding the endocrinology of the thyroid–pituitary axis and to describe in detail how the function of the thyroid gland can be disrupted by various xenobiotics. The rat thyroid gland appears to be particularly sensitive to adverse effects on these compounds and, inevitably, in long-term oncogenicity studies, results in the development of thyroid follicular cell tumours. At this time, ionizing radiation is the only acknowledged human thyroid carcinogen, a finding well established in experimental systems also. Although humans respond to goitrogenic stimuli as do animals, with the development of cellular hypertrophy, hyperplasia, and sometimes nodular lesions, even in its moderate to severe form, it is not an established aetiological factor for human thyroid cancer.

Because of marked species differences in thyroid gland physiology and apparent susceptibility to hypothyroidism, the rodent is an inappropriate model for the extrapolation of thyroid cancer risk to man for chemicals that operate secondary to hormone imbalance. For such compounds the onus is on the investigator to demonstrate this imbalance and to establish the cause and mechanisms of response. It is also essential to demonstrate that the compound is not genotoxic and has no effect other than through modifying thyroid homeostasis.

References

ALVARES, A. P., BICKERS, D. R. & KAPPAS, A. (1973) Polychlorinated biphenyls: a new type of inducer of cytochrome P-488 in the liver. *Proceedings of the National Academy of Science*, USA, **70**, 1321–5.

ANDERSON, R. R., NIXON, D. A. & AKASHA, M. A. (1988) Total and free thyroxine and triiodothyronine in blood serum of mammals. *Comparative Biochemistry and Physiology*, **89A**, 401–4.

ATTERWILL, C. K., COLLINS, P., BROWN, G. G. & HARLAND, R. F. (1987) The perchlorate discharge test for examining thyroid function in rats. *Journal of Pharmacological Methods*, **18**, 199–203.

ATTERWILL, C. K. & FOWLER, K. F. (1990) A comparison of cultured rat FRTL-5 and porcine thyroid cells for predicting the thyroid toxicity of xenobiotics. *Toxicology in Vitro*, **4**, 369–74.

ATTERWILL, C. K., JONES, C. & BROWN, C. G. (1992) Thyroid gland II – Mechanisms of species-dependent thyroid toxicity, hyperplasia and neoplasia induced by xenobiotics. In Atterwill, C. K. and Flack, J. D. (eds), *Endocrine Toxicology*, pp. 137–82. Cambridge: Cambridge University Press.

AXELRAD, A. A. & LEBOND, C. P. (1955) Induction of thyroid tumours in rats by a low iodine diet. *Cancer*, **8**, 339–67.

BARTER, R. A. & KLAASSEN, C. D. (1994) Reduction of thyroid hormone levels and alteration of thyroid function by four representative UCP-glucuronosyltransferase inducers in rats. *Toxicology and Applied Pharmacology*, **128**, 9–17.

BASTOMSKY, C. H. (1973) The biliary excretion of thyroxine and its glucuronic acid conjugate in normal and Gunn rats. *Endocrinology*, **92**, 35–40.

BIELSCHOWSKY, F. (1953) Chronic iodine deficiency as cause of neoplasia in thyroid and pituitary of aged rats. *British Journal of Cancer*, **7**, 203–13.

BOROWSKI, G. D., GAROFANO, C. D., ROSE, L. I., SPIELMAN, S. R., ROTMENSCH, H. R., GREENSPAN, A. M. & HOROWITZ, L. N. (1985) Effect of long-term amiodarone therapy on thyroid hormone levels and thyroid function. *American Journal of Medicine*, **78**, 443–50.

BOWMAN, W. C. & RAND, M. J. (1980) *Textbook of Pharmacology*. Oxford: Blackwell.

BROWN, C. G. (1987) Application of thyroid cell culture to the study of thyrotoxicity. In Atterwill, C. K. and Steele, C. E. (eds), *In Vitro Methods in Toxicology*, p. 165. Cambridge: Cambridge University Press.

BROWN, C. G., HARLAND, R. F., MAJOR, I. R. & ATTERWILL, C. K. (1987) Effects of toxic doses of novel histamine (H_2) antagonist on the rat thyroid gland. *Food and Chemical Toxicology*, **25**, 787–94.

CAPEN, C. C. (1983) Chemical injury of thyroid: pathologic and mechanistic considerations. In *Toxicology Forum*: 1983 Annual Winter Meeting, pp. 260–73. Washington: Bowers Reporting Company.

CAPEN, C. & MARTIN, S. (1989) The effects of xenobiotics on the structure and function of thyroid follicular and C cells. *Toxicologic Pathology*, **17**, 266–93.

CAVALIERI, R. R. & PITT-RIVERS, R. (1981) The effects of drugs on the distribution and metabolism of thyroid hormones. *Pharmacological Review*, **33**, 55–80.

COLLINS, W. T., JR., CAPEN, C. C., KASZA, L., CARTER, C. & DAILEY, R. F. (1977) Effects of polychlorinated biphenyl (PCB) on the thyroid gland of rats. *American Journal of Pathology*, **89**, 119–36.

DAVIES, D. T. (1993) Assessment of rodent thyroid endocrinology: advantages and pit-falls. *Comparative Haematology International*, **3**, 142–52.

DOHLER, K.-D., WONG, C. C. & VON ZUR MUHLEN, A. (1979) The rat as a model for the study of drug effects on thyroid function: consideration of methodological problems. *Pharmacological Therapeutics*, **5**, 305–18.

DONIACH, I. (1970) Aetiological consideration of thyroid carcinoma. In Smithers, D. (ed.), *Neoplastic Diseases at Various Sites: Tumours of the Thyroid Gland*. Edinburgh/London: E & S Livingston.

FURTH, J. (1954) Morphologic changes associated with thyrotropin-secreting pituitary tumours. *American Journal of Pathology*, **30**, 421–63.

GALTON, V. A. (1968) The physiological role of thyroid hormone metabolism. In James, V. H. T. (ed.), *Recent Advances in Endocrinology*, 8th Edn, pp. 181–206. London: Churchill.

GOLDMAN, M., PEASLEE, M. H. & NABER, S. P. (1970) The action of DDT on the iodide concentrating mechanism of the thyroid gland in male Sprague–Dawley rats. *American Zoologist*, **10**, 301–2.

GREAVES, P. (1990) *Histopathology of Preclinical Toxicity Studies: Interpretation and Relevance in Drug Safety Evaluation*. Oxford: Elsevier.

GRIESBACH, W. E., KENNEDY, T. H. & PURVES, H. D. (1945) Studies on experimental goitre. VI. Thyroid adenomata in rats on Brassica seed diets. *British Journal of Experimental Pathology*, **26**, 18–24.

GRUBECK-LOEWENSTEIN, B., BUCHAN, G., SADEGHI, R., KISSONERGHIS, M., LONDEI, M., TURNER, M., PIRICH, K., ROKA, R., NIEDERLE, B., KASSAL, H., WALDHAUSAL, W. & FELDMAN, M. (1989) Transforming growth factor β regulates thyroid growth. *Journal of Clinical Investigation*, **83**, 764–70.

HASEMAN, J. K., HUFF, J. & BOORMAN, G. A. (1984) Use of historical control data in carcinogenicity studies in rodents. *Toxicologic Pathology*, **12**, 126–35.

HAYDEN, D. W., WADE, G. G. & HANDLER, A. H. (1978) Goitrogenic effect of 4,4′-oxydianiline in rats and mice. *Veterinary Pathology*, **15**, 649–62.

HAYNES, R. C., JR. & MURAD, F. (1985) Thyroid and antithyroid drugs. In Gilman, A.

G., Goodman, L. S., Rall, T. W. and Murad, F. (eds), *The Pharmacological Basis of Therapeutics*, 7th Edn, pp. 1389–411. New York: Macmillan.

Hiasa, Y., Kitahori, Y., Oshima, M., Fujita, T., Yuasa, T., Konishi, N. & Miyashiro, A. (1982) Promoting effects of phenobarbital and barbital on development of thyroid tumours in rats treated with N-bis(2-hydroxypropyl)nitrosamine. *Carcinogenesis*, **3**, 1187–90.

Hill, R. N., Erdreich, L. S., Paynter, O. E., Roberts, P. A., Rosenhal, S. L., Wilkinson, C. F. (1989) Thyroid follicular cell carcinogenesis. *Fundamental and Applied Toxicology*, **12**, 629–97.

Huff, J. E., Eustis, S. L. & Haseman, J. K. (1989) Occurrence and relevance of chemically induced benign neoplasms in long-term carcinogenicity studies. *Cancer Metastasis Review*, **8**, 1–22.

Kaneko, J. J. (1980) Thyroid function. In Kaneko, J. J. (ed.), *Clinical Biochemistry of Domestic Animals*, 3rd Edn, pp. 491–512. New York: Academic Press.

Kuzan, F. B. & Prahlad, K. V. (1975) The effect of 1,2,3,4,10,10-hexachloro-1,4,4a,5,8,8a-hexahydroxyendo,exo-5,8-dimethionaphthalene (aldrin) and sodium ethylene-bisdithio-carbomate (Nabam) on the chick. *Poultry Science*, **54**, 1054–64.

Latropoulos, M. J. (1993/94) Endocrine considerations in toxicologic pathology. *Experimental Toxicologic Pathology*, **45**, 391–410.

Lueprasitsakul, W., Fang, S. L., Alex, S. & Braverman, L. E. (1991) Effect of the cardiac inotropic drug, OPC 8212, on pituitary-thyroid function in the rat. *Endocrinology*, **128**, 2709–14.

Mannering, G. J. (1971) In LaDu, B. N., Mandel, H. G. & Way, E. L. (eds), *Fundamentals of Drug Metabolism and Drug Disposition*, Baltimore: Williams & Wilkens.

McClain, R. M. (1989) The significance of hepatic microsomal enzyme induction and altered thyroid function in rats: implications for thyroid gland neoplasia. *Toxicologic Pathology*, **17**, 294–306.

McClain, R. M. (1992) Thyroid gland neoplasia: non-genotoxic mechanisms. *Toxicology Letters*, **64/65**, 397–408.

McClain, R. M., Levin, A. A., Posch, R. C. & Downing, J. C. (1989) The effect of phenobarbital on the metabolism and excretion of thyroxine in rats. *Toxicology and Applied Pharmacology*, **99**, 216–28.

McClain, R. M., Posch, R. C., Bosakowski, T. & Armstrong, J. M. (1988) Studies on the mode of action for thyroid gland tumour promotion in rats by phenobarbital. *Toxicology and Applied Pharmacology*, **94**, 254–65.

McTiernan, A. M., Weiss, N. S. & Dalling, J. R. (1984) Incidence of thyroid cancer in women in relation to previous exposure to radiation therapy and history of thyroid disease. *Journal of the National Cancer Institute*, **73**, 575–81.

Netter, F. H. (1974) Endocrine system and selected metabolic diseases, Volume 4. In Forsham, P. H. (ed.), *The CIBA Collection of Medical Illustrations*. New York: R. R. Donnelley & Sons.

Oppenheimer, J. H., Bernstein, G. & Surks, M. I. (1968) Increased thyroxine turnover and thyroidal function after stimulation of hepatocellular binding of thyroxine by phenobarbital. *Journal of Clinical Investigation*, **47**, 1399–406.

Oppenheimer, J. H., Shapiro, H. C., Schwartz, H. L. & Surks, M. I. (1971) Dissociation between thyroxine metabolism and hormonal action in phenobarbital-treated rats. *Endocrinology*, **88**, 115–19.

Oshima, M. & Ward, J. M. (1984) Promotion of N-methyl-N-nitrosourea-induced thyroid tumours by iodine deficiency in F334/NCr rats. *Journal of the National Cancer Institute*, **73**, 289–96.

OSTP, Office of Science and Technology Policy (1985) Chemical carcinogens: a review of the science and its associated principles. *Federal Register*, **50**, 10371–442.

PAYNTER, O. E., BURIN, G. J., JAEGER, R. B. & GREGARIO, C. A. (1986) *Neoplasia Induced by Inhibition of Thyroid Gland Function* (Guidance for Analysis and Evaluation). Hazard Evaluation Division, U.S. Environmental Protection Agency, Washington, D.C.

(1988) Goitrogens and thyroid follicular cell neoplasia. Evidence for a threshold process. *Regulatory Toxicology and Pharmacology*, **8**, 102–19.

PENDERGRAST, W. J., MILMORE, B. K. & MACUS, S. C. (1961) Thyroid cancer and thyrotoxicosis in the United States. Their relationship with endemic goiter. *Journal of Chronic Disease*, **13**, 22–38.

POTTER, C. L., SIPES, I. G. & RUSSEL, D. H. (1983) Hypothyroxinemia and hypothermia in rats in responses to 2,3,7,8-tetrachlorodibenzo-p-dioxin administration. *Toxicology and Applied Pharmacology*, **69**, 89–95.

READER, S. J. (1987) Assessment of the biopotency of antithyroid drugs and using porcine thyroid cells. *Biochemistry and Pharmacology*, **36**, 1825–8.

REFETOFF, S., ROBIN, N. I. & FANG, U. S. (1970) Parameters of thyroid function in serum of 16 selected vertebrate species: a study of PBI, serum T_4, free T_4 and the pattern of T_4 and T_3 binding to serum protein. *Endocrinology*, **86**, 793–805.

RON, E., KLEINERMAN, R. A., BOICE, J. D., LIVOLSI, V. A., FLANNERY, J. T. & FRAUMENI, J. F. (1987) A population-based case-control study of thyroid cancer. *Journal of the National Cancer Institute*, **79**, 1–12.

RYAN, D., LU, A. Y. H. & LEVIN, W. (1978) Purification of cytochrome P-450 and P-448 from rat liver microsomes. *Methods in Enzymology*, **52**, 117–23.

SIMESEN, M. G. (1980) Calcium, phosphorus and magnesium metabolism. In Kaneko, J. J. (ed.) *Clinical Biochemistry of Domestic Animals*, 3rd Edn, pp. 575–648. New York: Academic Press.

SMITH, P., WYNFORD-THOMAS, D., STRINGER, B. M. J. & WILLIAMS, E. D. (1986) Growth factor control of rat thyroid follicular cell proliferation. *Endocrinology*, **119**, 1439–45.

SRINIVASAN, V., MOUDGAL, M. R. & SARAMA, P. S. (1957) Studies of goitrogenic agents in food. I. Goitrogenic action of groundnut. *Journal of Nutrition*, **61**, 87–95.

STRINGER, B. M. J., WYNFORD-THOMAS, D. & WILLIAMS, E. D. (1985) *In vitro* evidence for an intracellular mechanism limiting the thyroid follicular cell growth response to thyrotropin. *Endocrinology*, **116**, 611–15.

TAKAOKA, M., MANABE, S., YAMOTO, T., TERANISHI, M., MATSUNUMA, N., MASUDA, H. & GOTO, N. (1994) Comparative study of goitrogenic actions of 3-substituted 1,2,4-triazoles in rats. *Journal of Veterinary Medical Science*, **56**, 341–6.

TAUROG, A. (1970) Thyroid peroxidase and thyroxine biosynthesis. *Recent Progress in Hormone Research*, **26**, 189–247.

THOMAS, G. A., WILLIAMS, D. & WILLIAMS, E. D. (1991) Reversibility of malignant phenotype in monoclonal thyroid in the mouse. *British Journal of Cancer*, **63**, 213–16.

VAN ETTEN, C. H. (1969) Goitrogens. In Liener, I. E. (ed.), *Toxic Constituents of Plant Foodstuffs*, pp. 103–42. New York: Academic Press.

WENZEL, K. W. (1981) Pharmacological interference with *in vitro* tests for thyroid function. *Metabolism*, **30**, 717–32.

WILLIAMS, D. W., WILLIAMS, E. D. & WYNFORD-THOMAS, D. (1988) Loss of dependence on IGF-1 for proliferation of human thyroid adenoma cells. *British Journal of Cancer*, **57**, 535–9.

WOO, Y.-T., LAI, D. Y., ARGOS, J. C. & ARGOS, M. F. (1985) Chemical induction of cancer: structural bases and biological mechanisms. In Woo, Y. *et al.* (eds), *Aliphatic Polyhalogenated Carcinogens*, Vol. 3B, pp. 357–94. New York: Academic Press.

WOODMAN, D. D. (1988) The use of chemical biochemistry for assessment of endocrine system toxicology. In Keller, P. & Bogin, E. (eds), *The Use of Clinical Biochemistry in Toxicologically Relevant Animal Models and Standardisation and Quality Control in Animal Biochemistry*, pp. 63–77. Basel: Hexagon-Roche.

WYNFORD-THOMAS, D., STRINGER, B. M. J. & WILLIAMS, E. D. (1982a) Goitrogen-induced thyroid growth in the rat: a quantitative morphometric study. *Journal of Endocrinology*, **94**, 131–40.

(1982b) Dissociation of growth and function in the rat thyroid during prolonged goitrogen administration. *Acta Endocrinology*, **101**, 210–16.

(1982c) Desensitisation of rat thyroid to the growth-stimulating action of TSH during prolonged goitrogen administration. *Acta Endocrinology*, **101**, 562–9.

Section 8.3 Other Endocrine Organs

G. O. EVANS

Tests for assessing the functionality and injury to several of the endocrine glands other than the thyroid glands will be discussed in this section. It is important to recognize the general frequency of toxic findings in these organs (see Section 8.1) and to recognize the role of reproductive and teratological studies in the detection of toxicants to the reproductive systems.

8.3.1 Adrenals

The adrenal gland has several distinct anatomical zones which also differ functionally. The outer adrenal cortex has three distinct zones: the outer zona glomerulosa, the intermediate zona fasciculata and the inner zona reticularis which adjoins the adrenal medulla (Colby, 1987a). The zona glomerulosa is the site of mineralocorticoid production which includes aldosterone, and this zone is therefore associated with electrolyte homeostasis. The inner two cortical zones secrete glucocorticoids, adrenal androgens and oestrogens. These zonal differences are related to two steroidogenic enzyme systems: the mitochondrial $P450_{cmo}$ of the zona glomerulosa and the microsomal $P450_{17\alpha}$ of the inner zones. The biosynthesis and secretion of these adrenal cortical hormones are influenced by the pituitary trophic hormone, ACTH (adrenocorticotrophic hormone; Colby, 1987a) and its associated feedback mechanisms (Figure 8.3.1). These glucocorticoids influence the metabolism of carbohydrates, lipids and proteins in the liver, muscle and adipose tissues; thus, severe toxic effects on the adrenals cause metabolic changes which may be detected by measurements of plasma glucose, cholesterol and proteins. Corticosterone, cortisol, aldosterone and catecholamines are the primary adrenal hormones which are of interest in toxicology studies.

Toxic effects on the adrenal cortex involving steroid metabolism can be caused by several different mechanisms (Ribelin, 1984) which include alterations of cholesterol synthesis, conversion of cholesterol to 5-pregnenolone, or 11β-hydroxylation (Figure 8.3.2). Several adrenal cortical toxins were reviewed by Szabo and Lippe (1989). Some xenobiotics may alter ACTH secretion via the

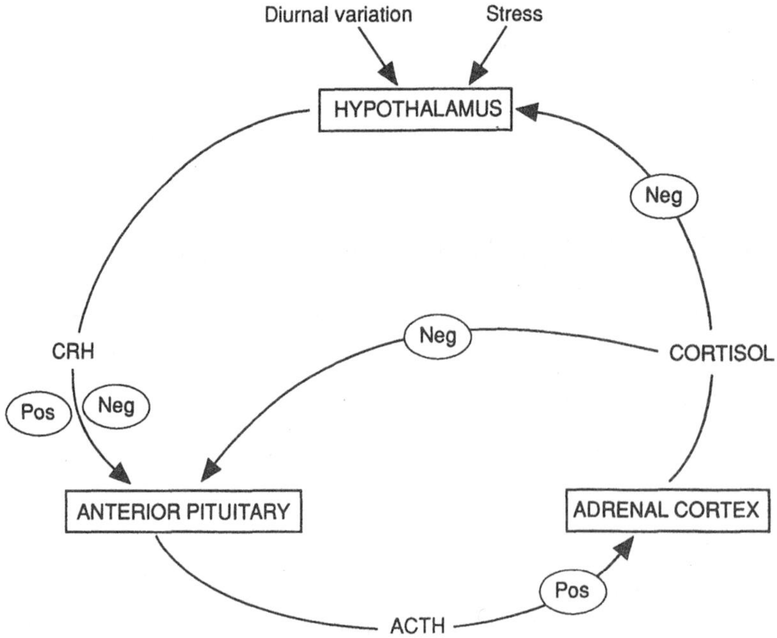

Figure 8.3.1 Simplified pathways for biosynthesis of steroid hormones

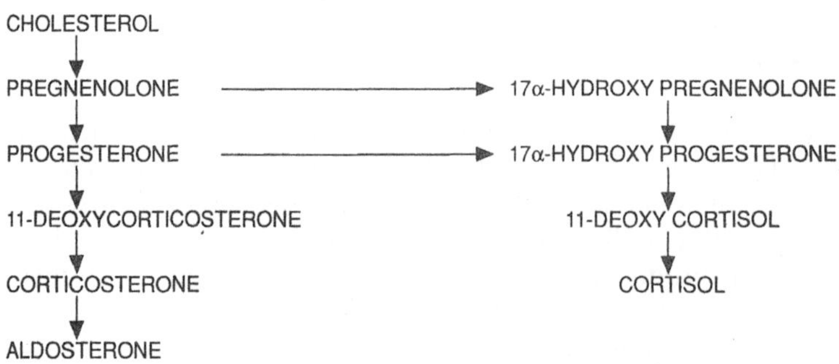

Figure 8.3.2 Metabolism of glucocorticoids and mineralocorticoids

pituitary and therefore affect the rate of adrenal steroid secretion, while other compounds acting on membrane receptors, protein synthesis etc. will alter steroidogenesis. For example, spironolactone competes for aldosterone receptors in the kidney although it has extra-renal sites of action (Colby and Longhurst, 1992). Several compounds that inhibit adrenal gland steroidogenesis can also affect gonadal steroidogenesis, particularly in the testes (Thomas and Keenan, 1986).

The steroid hormones are sparingly soluble in aqueous media, and are transported in plasma mainly bound to proteins, but it is the unbound or free hormone levels that are important for the biological activity in the target tissues. Corticosteroids are bound to plasma proteins including transcortin (or corticosteroid binding globulin, CBG) and albumin; these proteins have differing affinities and binding capacities for the corticoids. Transcortin is synthesized in the liver and it appears to be present in most species (Seal and Doe, 1966; Westphal, 1971), but its functions are not well defined. As a consequence of hepatic toxicity or excessive renal loss of protein, some hormonal alterations can be caused by changes in plasma binding proteins.

The mineralocorticoids influence electrolyte transport by regulating renal reabsorption and excretion of sodium, chloride, hydrogen and potassium ions thereby affecting blood pressure homeostasis. The primary mineralocorticoid is aldosterone and its effects are dependent upon the renin-angiotensin system, plasma potassium and/or sodium concentrations. Major changes due to adrenal toxicity, therefore, can be detected by changes in electrolyte balance without resorting to hormone measurements. The control mechanisms for the glucocorticoids and mineralocorticoids are essentially independent of each other, but both may be affected in severe adrenal toxicy.

The inner adrenal medulla secretes both epinephrine (adrenaline) and norepinephrine (noradrenaline) in response to sympathetic nerve stimulation (Colby, 1987b). Both of these hormones influence a wide range of metabolic activities, smooth muscle functions and glandular secretions. Less is known about the toxic effects of chemicals on medullary function with relatively few reports in the literature in comparison to adrenal cortical toxins. Chemically induced stress can provoke rapid responses in the secretion of catecholamines, which have relatively short biological half-lives, contrasting with toxicity due to xenobiotics which generally require longer periods of repeated exposure.

8.3.1.1 Factors Affecting Adrenal Hormone Assays

One of the main differences between species is the dominant glucocorticoid, which in rats and mice is corticosterone, but in dogs, rabbits, non-human primates and man is cortisol. As the adrenal gland:body weight ratio is relatively large and the primary glucocorticoid is cortisol in the guinea pig, this species is a useful model for assessing adrenal toxicity (Colby, 1988).

The circulating levels of adrenal cortical hormones in rats depend on several factors including age, season and strain (Kuhn *et al.*, 1983; Wong *et al.*, 1983), sex and light (Critchlow *et al.*, 1963), handling and environment (Barrett and Stockham, 1963; Dunn and Scheving, 1971; Dohler *et al.*, 1978; Jurcovicova *et al.*, 1984) and circadian rhythm (D'Agostino *et al.*, 1982). In dogs, a circadian rhythm is not apparent (Johnston and Mather, 1978) but episodic secretion of cortisol and ACTH has been reported (Kemppainen and Sartin, 1984).

Measurements of basal plasma cortisol following administration of ACTH (or synthetic analogues) can be used as a functional test for the adrenals (Garnier *et al.*, 1990). Another functional test involves ACTH measurements after the administration of metyrapone, which inhibits adrenal 11β-hydroxylase activity, and causes a transient reduction in cortisol synthesis and increased secretion of 11-deoxycortisol (Orth *et al.*, 1988). Urinary corticoid measurements (in 24 h but not short-term samples) can be used to assess adrenocortical function as alternatives to blood-sampling techniques (Hilfenhaus, 1977).

A circadian rhythm has been demonstrated for plasma aldosterone in the rat

(Gomez-Sanchez *et al.*, 1976) and in the rabbit (Vernay *et al.*, 1984). Dietary sodium can also influence levels of plasma aldosterone (Kotchen *et al.*, 1983). Urinary aldosterone measurements are of little value as less than 1% of secreted aldosterone is excreted unchanged via the kidneys.

Problems in collecting suitable basal blood samples from rodents may be avoided by using in-dwelling catheters, as other blood-sampling methods produce elevated values for epinephrine, norepinephrine and catecholamines (Kvetnansky *et al.*, 1978; Popper *et al.*, 1978).

8.3.2 Gonads

Chemicals which are toxic to the reproductive system may act directly because they possess a structural similarity to an endogenous compound (e.g. antagonists of endogenous hormones, oral contraceptives) or because of a general chemical action (e.g. alkylating agents). Some toxins act indirectly by altering enzymes involved in steroid synthesis and in hepatic metabolism, or via the neuroendocrine centres by influencing the gonadotrophic hormones (Mattison and Thomford, 1989). Other compounds act on the meiotic and mitotic stages of spermatogenesis. Sometimes, as with other fields of toxicology, the cause of the toxicity may not be the most obvious candidate; the toxicity observed with dogs given oestrogens is due to bone marrow suppression and not by oestrogenic action on the gonads (Johnson, 1989).

Gonadectomy may be a useful adjunct in elucidating the mechanisms of testicular or ovarian toxicity. By removal of a particular endocrine organ, it is possible to establish the role of that organ in the development of the endocrine effect and the relationships to feedback mechanisms.

8.3.3 Testes

The release of several peripheral hormones is controlled by the secretion of trophic hormones from the pituitary, which in turn are regulated by releasing hormones secreted by the hypothalamus. Some of these hormones are listed in Table 8.3.1. Under normal circumstances the circulating hormones regulate the output of releasing and trophic hormones by feedback mechanisms.

Whilst the weights of accessory glands, testes and epididymes are the primary indicators of a possible alteration in androgen status, measurement of circulating hormones and some proteins and enzymes can assist in the diagnosis of testicular toxicity. The mammalian testis can be functionally separated into the seminiferous tubules and interstitial region with its Leydig cells. The major function of the Leydig cells is maintenance of spermatogenesis which is partially mediated by the production of testosterone and dihydrotestosterone. In the seminiferous tubules, there are two major cell types: the spermatogonial and the Sertoli cells. The functions of the Sertoli cells include the synthesis of androgen binding proteins, nutrition for the developing sperm cells and maintenance of the blood–testis barrier (Gangolli and Phillips, 1993).

The endocrine control of testicular function is mainly exercised by the gonadotrophic luteotrophin (LH) and follitrophin (FSH). The primary target cells

Table 8.3.1 Major functions of pituitary hormones

Hormone	Function
Adenohypophysis	
ACTH	Stimulates production of glucocorticosteroids by adrenal cortical cells.
FSH	Stimulates ovarian follicle growth and spermatogenesis.
GH	Accelerates tissue growth.
LH	Induces follicular maturation, ovulation, formation of corpus luteum and oestrogen secretion. Stimulates androgen secretion in males.
PRL	Promotes mammary gland development and lactation. Regulates Leydig cell functions.
TSH	Regulates thyroid hormone synthesis.
MSH	Induces melanin synthesis.
Neurohypophysis	
ADH	Promotes water retention.
Oxytocin	Affects lactation and uterine function during parturition.

of LH and FSH are the Leydig and Sertoli cells respectively, but the homeostatic control of this hypothalamo–pituitary–testicular (HPT) axis is complex, with several overlapping functions and feedback mechanisms at different sites (Sharpe, 1982; Figure 8.3.3). Prolactin appears to act synergistically with LH in regulating testosterone metabolism.

Elevations of gonadotrophins can indicate testicular injury, with an increase of plasma LH indicating Leydig cell dysfunction and elevated plasma FSH indicative of damage to seminiferous tubules, although this is not always a clear-cut finding.

Rats exhibit circadian rhythms for plasma testosterone (Kalra and Kalra, 1977) and episodic fluctuations of this hormone have been reported in rabbits (Moor and Younglai, 1975), dogs (de Palatis *et al.*, 1978) and rhesus monkeys (Plant, 1981). LH is secreted in a pulsatile manner and these pulses can cause plasma levels to vary by more than twofold baseline values (Santen and Bardin, 1973). Given these variations, blood collection procedures must be carefully controlled if meaningful results are to be obtained. Testosterone may be measured by radioimmunochemical, gas-liquid chromatographic or fluorimetric assays; the other hormones can be measured by several different immunoassays or chemical methods (Edqvist and Stabenfeldt, 1989).

Testosterone-oestradiol binding globulin (TEBG) or sex hormone binding globulin (SHBG) is a glycoprotein of hepatic origin which shows a greater affinity for androgens than oestrogens. Androgen binding proteins are produced by the Sertoli cells and include androgenic binding glycoprotein (ABP) which has a high affinity for testosterone and dihydrotestosterone, and has been measured in studies of seminiferous tubular function (Gunsalus *et al.*, 1978; Gunsalus *et al.*, 1981; Spitz

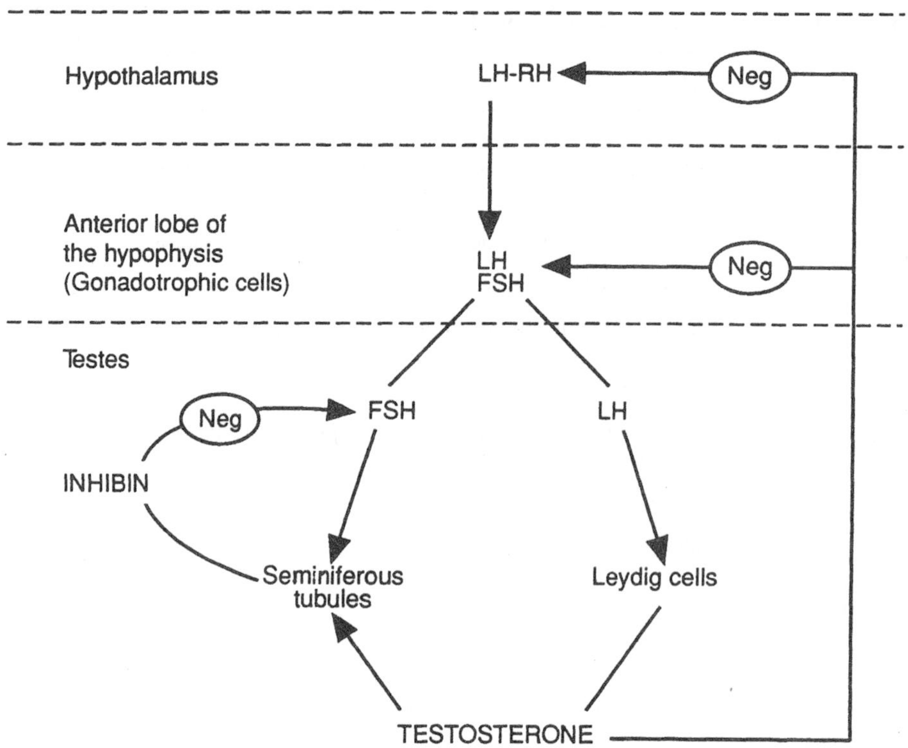

Figure 8.3.3 Schematic diagram for hormonal regulation in males

et al., 1985). The interrelationship of these androgen binding proteins is uncertain (Bardin *et al.*, 1981; Cheng *et al.*, 1985).

Testis-specific enzymes occur and the isoenzyme of lactate dehydrogenase (LDH), referred to as LDH-X or LDH-C4, is the most widely studied. The synthesis of LDH-C4 isoenzyme occurs at a particular stage of germ cell development (Meistrich *et al.*, 1977; Wheat *et al.*, 1977) and elevations of plasma LDH-C4 have been reported in several studies of acute testicular toxicity in the rat, where it is not normally detected in circulating blood (Haqqi and Adhami, 1982; Itoh and Ozasa, 1985; Reader *et al.*, 1991). Recent work with several testicular toxicants has shown that creatinuria can be used as a marker of testicular damage, with creatine appearing to be more sensitive than plasma LDH-C4, testosterone or testis mass in the detection of damage (Moore *et al.*, 1992; Timbrell *et al.*, 1994).

8.3.4 Ovaries

Hormonal control of the female reproductive system involves several hormones of the hypothalamic–pituitary–ovarian axis. The hypothalamic secretion of a decapeptide, gonadotrophin releasing hormone (GnRH), acts in a pulsatile manner to stimulate the release of FSH and LH from the anterior pituitary gland. These two gonadotrophins, FSH and LH, are dimeric glycoproteins of similar

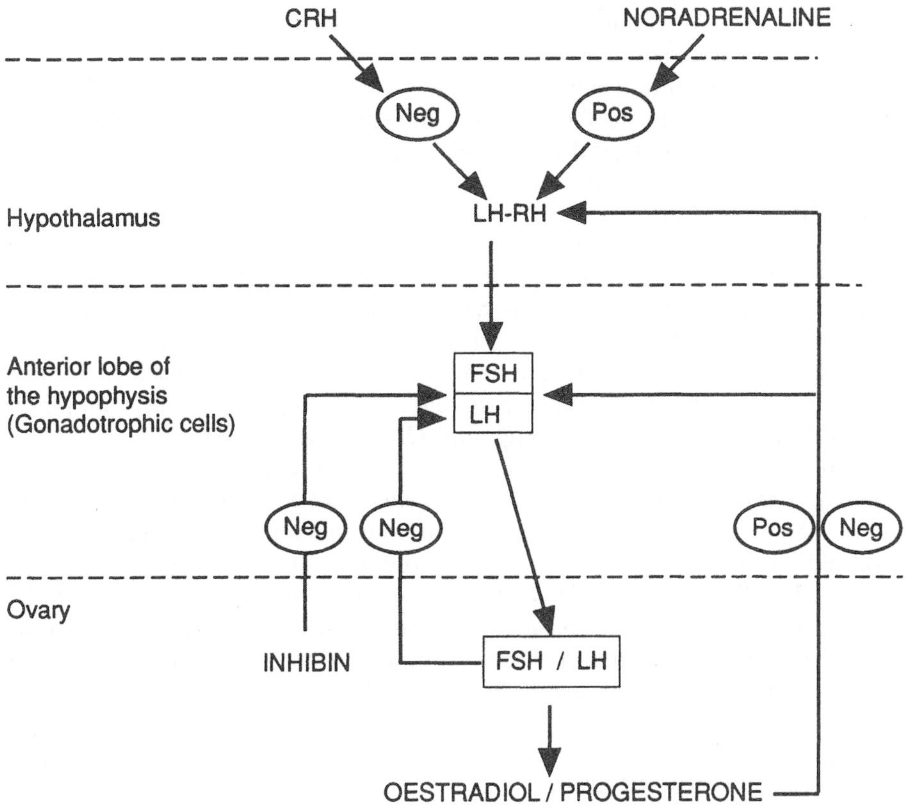

Figure 8.3.4 Schematic diagram for hormonal regulation in females

molecular mass, and they act on specific ovarian receptors to stimulate production of steroid hormones. The targets for LH are the theca cells, corpus luteum, promotion of ovulation and differentiation of luteal cells (luteinization). FSH controls maturation of the follicles and stimulates steroid hormone production by the granulosa cells (Wilson and Leigh, 1992; Gangolli and Phillips, 1993; Figure 8.3.4). Of these steroid hormones, 17β-oestradiol is the most potent oestrogenic hormone. Several positive and negative feedback mechanisms operate within this axis.

Other hormones involved in the female reproductive system include androgens, gestagens and prolactin. The androgens modulate production and secretion of the hypophyseal gonadotrophins. Gestagens include the steroid progesterone which is synthesized in the corpus luteum and placenta. Progesterone acts in the uterus to control the secretory transformation of the endometrium. Prolactin is a single chain polypeptide, and its hypophyseal synthesis and release are controlled by neuronal dopamine (prolactin inhibiting factor, PIF) and prolactin releasing factor (PAF).

Xenobiotics may inhibit ovulation, ovum transport, fertilization or implantation. The toxicity may result from a direct action because of structural similarity with endogenous hormones, e.g. oral contraceptives, and anti-oestrogens which alter the normal feedback inhibition control of oestrogen synthesis, in turn increasing secretion of LH and FSH releasing hormones. Alternatively, xenobiotics may show a wide range of indirect actions such as alteration of ovarian microsomal mono-oxygenase activities (Haney, 1985; Wilson and Leigh, 1992; Ratcliffe *et al.*, 1993). Chemicals acting on the hypothalamic–pituitary–ovarian axis can inhibit release of gonadotrophins and thus cause a secondary effect on ovarian steroid synthesis. Changes in oestrogenic metabolism may also cause alterations of immune functions (Luster *et al.*, 1985).

Ovulatory cycles are highly variable in the different species, and the several rhythmic patterns of the plasma reproductive hormones together with natural episodic changes can cause problems when interpreting results from these assays (Kreiger, 1979; see Section 8.1). Further complications include reproductive senescence, e.g. in rodents where female rats develop irregularities in their oestrous cycle at about ten months of age, and temporal changes in plasma hormone levels induced by test compound (Jones *et al.*, 1987).

Although some studies have shown correlation between mammary gland changes and prolactin levels, in rodents the results may be sometimes disappointing because of stress and other effects: serial inter-animal plasma measurements performed in additional studies may be required to demonstrate significant changes. LH, FSH and prolactin can be measured by immunoassays: for both LH and FSH, there appears to be considerable cross-reactivity between several species which enables measurements of these hormones to be performed, e.g. rat FSH reagents can be used to measure FSH in mice, hamsters and guinea pigs.

8.3.5 Hypothalamus–Pituitary

The adenohypophysis secretes several peptide hormones which include somatotrophin (GH), prolactin (PRL), thyroid stimulating hormone (TSH), LH, FSH, adrenocorticotrophic hormone (ACTH) and melanocyte stimulating hormone (MSH). The neurohypophysis secretes antidiuretic hormone (ADH or vasopressin) and oxytocin (Reichlin, 1985). The major functions of these pituitary hormones are shown in Table 8.3.1.

In the previous sections on thyroidal and gonadal hormones, some of the roles for pituitary hormones have been described where the hypothalamic release-stimulating hormones together with the pituitary secretion of trophic hormones and feedback mechanisms play an important role in hormonal homeostasis for several potential target organs. The combination of thyroidal and gonadal toxicity findings in a study must lead to a thorough examination of hypothalamic–pituitary function. (This expression of pituitary toxicity via other endocrine organs has probably contributed to the paucity of published data for specific effects on the pituitary gland.) Pituitary toxicity may develop as a consequence of changes in negative feedback mechanisms.

8.3.6 Calcium Homeostasis

The major hormones involved in calcium homeostasis are parathyroid hormone (PTH; parathyrin: parathormone), calcitonin and 1,25-dihydroxyvitamin D (calcitriol). As a general rule, effects on the parathyroid function will be detected from histopathological examination of these glands, and it is not necessary to perform these hormone assays in the vast majority of studies (Capen, 1983; Capen and Rosol, 1989). Although some immunoassays are available for PTH, there are species differences and some reagents measure inactive fragments of PTH in addition to the intact hormone. Assays which measure the amino- or N-terminal region of the polypeptide PTH are more suitable than assays for the carboxy- or C-peptide fragment, which is biologically inactive. PTH and calcitonin are both measurable by radioimmunoassay (Selby and Adams, 1994). In rats, PTH values vary with age (Kalu *et al.*, 1983; Kalu and Hardin, 1984), and nutritional status (Talmage *et al.*, 1975)

Parathyroid toxicity can sometimes be detected by measuring plasma calcium which is usually measured as total calcium. As approximately 30 to 40% of plasma calcium is bound to albumin with some binding to the globulin fractions, interpretation of plasma total calcium levels should always allow for variations in plasma proteins, or measurements of ionized or unbound plasma calcium fractions should be considered.

Toxic injury to the parathyroids following acute or chronic administration of xenobiotics occurs rarely, but parathyroid function may be altered by a wide variety of chemicals which increase or decrease plasma ionic calcium levels. This secondary effect on the parathyroids can become more complicated in certain situations: for example, hypercalcaemia in acute renal failure may be observed in secondary hyperparathyroidism as a consequence of PTH stimulation following the initial hypocalcaemia.

8.3.7 Summary

The application of hormone measurements is to some extent limited by the lack of specific assays and by the considerable physiological inter- and intra-animal variations which occur; additionally, there are some marked differences between species. Knowledge of the different mechanisms which can produce toxicity is expanding through the use of plasma hormone measurements particularly for some endocrine organs, e.g. the thyroid. However, the diversity of the mechanisms for indirectly acting toxins and the complex physiological feedback mechanisms mean that several hormone measurements currently do little to extend the histopathological findings for many endocrinological toxins. Nevertheless, measurements of circulating levels of one or more hormones together with other specific indicators of endocrine organ function are useful tools in defining the level at which functional impairment occurs, the progression of lesions and distinguishing between alterations of function from effects due to toxic injury (Capen and Martin, 1989).

References

BARDIN, C. W., MUSTO, N., GUNSALUS, G. L., KOTITEN, N., CHENG, S. L., LARREA, F. & BECKER, R. (1981) Extracellular androgen binding proteins. *Annual Reviews of Physiology*, **43**, 189–98.

BARRETT, A. M. & STOCKHAM, M. A. (1963) The effect of housing conditions and simple experimental procedures upon the corticosterone level in the plasma of rats. *Journal of Endocrinology*, **26**, 97–105.

CAPEN, C. C. (1983) Structural and biochemical aspects of parathyroid gland function in animals. In Jones, T. C., Mohr, U. & Hunt, R. D. (eds), *Endocrine System*, pp. 217–47. Berlin: Springer-Verlag.

(1989) The calcium regulating hormones parathyroid hormone, calcitonin and cholecaliferol. In McDonald, L. E. (ed.), *Veterinary Endocrinology*, 4th Edn, pp. 92–185. Philadelphia: Lea & Febiger.

CAPEN, C. C. & MARTIN, S. L. (1989) Mechanisms that lead to disease of the endocrine system in animals. *Toxicologic Pathology*, **17**, 234–49.

CAPEN, C. C. & ROSOL, T. J. (1989) Recent advances in the structure and function of the parathyroid gland in animals and the effect of xenobiotics. *Toxicologic Pathology*, **17**, 333–56.

CHENG, C. Y., MUSTO, N. A., GUNSALUS, G. L., FRICK, J. & BARDIN, C. W. (1985) There are two forms of androgen binding proteins in human testes. *Journal of Biological Chemistry*, **260**, 5631–40.

COLBY, H. D. (1987a) The adrenal cortex. In Hedge, G. A., Colby, H. D. & Goodman, R. L. (eds), *Clinical Endocrine Physiology*, pp. 127–59. Philadelphia: W. B. Saunders.

(1987b) The adrenal medulla. In Hedge, G. A., Colby, H. D. & Goodman, R. L. (eds), *Clinical Endocrine Physiology*, pp. 297–315. Philadelphia: W. B. Saunders.

(1988) Adrenal gland toxicity: chemically induced dysfunction. *Journal of the American College of Toxicology*, **7**, 45–69.

COLBY, H. D. & LONGHURST, P. A. (1992) Toxicology of the adrenal gland. In Atterwill, C. K. & Flack, J. D. (eds), *Endocrine Toxicology*, pp. 243–58. Cambridge: Cambridge University Press.

CRITCHLOW, V., LIEBELT, R. A., BAR-SELA, M., MOUNTCASTLE, W. & LIPSCOMB, H. (1963) Sex difference in the resting pituitary-adrenal function of in the rat. *American Journal of Physiology*, **205**, 807–15.

D'AGOSTINO, J. B., VAETH, G. F. & HENNING, S. J. (1982) Diurnal rhythm of total and free concentrations of serum corticosterone in the rat. *Acta Endocrinology*, **100**, 85–90.

DE PALATIS, L., MOORE, J. & FALVO, R. E. (1978) Plasma concentrations of testosterone and LH in the male dog. *Journal of Reproduction and Fertility*, **52**, 201–7.

DOHLER, K.-D., WONG, C.-C., GAUDSSUHN, D., VON ZUR MUHLEN, A., GARTNER, K. & DOHLER, U. (1978) Site of blood sampling in rats as a possible source of error in hormone determinations. *Journal of Endocrinology*, **79**, 141–2.

DUNN, J. & SCHEVING, L. (1971) Plasma corticosterone levels in rats killed sequentially at the 'trough' or 'peak' of the adrenocortical cycle. *Journal of Endocrinology*, **49**, 347–8.

EDQVIST, L.-E. & STABENFELDT, G. H. (1989) Clinical reproductive endocrinology. In Kaneko, J. J. (ed.), *Clinical Biochemistry of Domestic Animals*, 4th Edn, pp. 650–77. Academic Press: San Diego.

GANGOLLI, S. D. & PHILLIPS, J. C. (1993) Assessing chemical injury to the reproductive system. In Anderson, D. & Conning, D. M. (eds), *Experimental Toxicology*, 2nd Edn, pp. 376–404. Cambridge: Royal Society of Chemistry.

GARNIER, F., BENOIT, E., VIRAT, M., OCHOA, R. & DELATOUR, P. (1990) Adrenal cortical response in clinically normal dogs before and after adaptation to a housing environment. *Laboratory Animals*, **24**, 40–3.

GOMEZ-SANCHEZ, C., HOLLAND, O. B., HIGGINS, J. R., KERN, D. C. & KAPLAN, N. M. (1976) Circadian rhythms of serum renin activity and serum corticosterone, prolactin and aldosterone concentrations in the male rat on normal and low-sodium diets. *Endocrinology*, **99**, 567–72.

GUNSALUS, G. L., LARREA, F., MUSTO, N. A., BECKER, R. R., MATHER, J. P. & BARDIN, C. W. (1981) Immunoassay of androgen binding protein in blood: a new approach for study of the seminiferous tubule. *Science*, **200**, 65–6.

GUNSALUS, G. L., MUSTO, N. A. & BARDIN, C. W. (1978) Androgen binding protein as a marker for Sertoli cell function. *Journal of Steroid Biochemistry*, **15**, 99–106.

HANEY, A. F. (1985) Effect of toxic agents on ovarian function. *Endocrine Toxicology, Target Organ Toxicology series*, pp. 181–210. New York: Raven Press.

HAQQI, T. M. & ADHAMI, U. M. (1982) Testicular damage and change in serum LDH isoenzyme patterns induced by multiple sub-lethal doses of apholate in albino rats. *Toxicology Letters*, **12**, 199–205.

HILFENHAUS, M. (1977) Urinary excretion of corticosterone as a parameter of adrenocortical function in rats. *Naunuyn Schmiedebergs Archives of Pharmacology*, Suppl. **11**, 297:R41.

ITOH, R. & OZASA, H. K. (1985) Changes in serum lactate dehydrogenase isozyme X activity observed after cadmium administration. *Toxicology Letters*, **28**, 151–4.

JOHNSON, A. N. (1989) Comparative aspects of contraceptive steroids – effects observed in beagle dogs. *Toxicologic Pathology*, **17**, 389–95.

JOHNSTON, S. D. & MATHER, E. C. (1978) Canine plasma cortisol (hydroxycortisone) measured by radioimmunoassay: clinical absence of diurnal variation and results of ACTH stimulation and dexamethasone suppression tests. *American Journal of Veterinary Research*, **39**, 1766–70.

JONES, M. K., WEISENBURGER, W. P., SIPES, I. G., RUSSELL, D. H. (1987) Circadian alterations in prolactin, corticosterone, and thyroid hormone levels and down-regulation of prolactin receptor activity by 2,3,7,8-tetrachlorodibenzo-*p*-dioxin. *Toxicology and Applied Pharmacology*, **87**, 337–50.

JURCOVICOVA, J., VIGAS, M., KLIR, P. & JEZOVA, D. (1984) Response of prolactin, growth hormone and corticosterone to morphine administration or stress exposure in Wistar-AVN and Long-Evans rats. *Endocrinology Experimentia*, **18**, 209–14.

KALRA, P. S. & KALRA, S. P. (1977) Circadian periodicities of serum androgens, progesterone, gonadotrophins and luteinizing hormone-releasing hormone in male rats: the effects of hypothalamic deafferentation, castration and adrenalectomy. *Endocrinology*, **101**, 1821–7.

KALU, D. N., COCKERMAN, R., YU, B. P. & ROOS, B. A. (1983) Lifelong dietary modulation of calcitonin levels in rats. *Endocrinology*, **113**, 2010–15.

KALU, D. N. & HARDIN, R. H. (1984) Age, strain and species differences in circulating parathyroid hormone. *Hormone and Metabolic Research*, **16**, 654–7.

KEMPPAINEN, R. J. & SARTIN, J. L. (1984) Evidence for episodic but not circadian rhythm in plasma concentrations of adreno-corticotrophin, cortisol and thyroxine in the dog. *Journal of Endocrinology*, **103**, 219–26.

KOTCHEN, T. A., GUTHRIE, G. P., GALLA, J. H., LUKE, R. & WELCH, W. J. (1983) Effects of NaCl on renin and aldosterone responses to potassium depletion. *American Journal of Physiology*, **244**, E164–9.

KREIGER, D. T. (1979) *Endocrine Rhythms*. California: Raven Press.

KUHN, E. R., BELLON, K., HUYBRECHTS, L. & HEYNS, W. (1983) Endocrine differences between the Wistar and Sprague–Dawley laboratory rat: influence of cold adaptation. *Hormone and Metabolism Research*, **15**, 491–8.

KVETNANSKY, R., SUN, C. L., LAKE, C. R., THOA, M., TORDA, T. & KOPIN, I. J. (1978) Effects of handling and forced immobilization on rat plasma levels of epinephrine, norepinephrine and dopamine-β-hydroxylase. *Endocrinology*, **103**, 1868–74.

LUSTER, M. I., PFEIFER, R. W. & TUCKER, A. N. (1985) Influence of sex hormones on immunoregulation with specific references to natural and environmental estrogens. In Thomas, J. A. (ed.), *Endocrine Toxicology, Target Organ Toxicology series*, pp. 67–85. New York: Raven Press.

MATTISON, D. R. & THOMFORD, P. J. (1989) The mechanism of action of reproductive toxicants. *Toxicologic Pathology*, **17**, 364–76.

MEISTRICH, M. L., TROSTLE, P. K., FRAPART, M. & ERICKSON, R. P. (1977) Biosynthesis and localization of lactate dehydrogenase X in pachytene spermatocytes and spermatids of mouse testes. *Developmental Biology*, **60**, 428–41.

MOOR, B. C. & YOUNGLAI, E. V. (1975) Variations in peripheral level of serum LH and testosterone in adult male rabbits. *Journal of Reproduction and Fertility*, **42**, 259–66.

MOORE, N. P., CREASY, D. M., GRAY, T. J. B. & TIMBRELL, J. A. (1992) Urinary creatine profiles after administration of cell-specific toxicants to the rat. *Archives of Toxicology*, **66**, 435–42.

ORTH, D. N., PETERSON, M. E. & DRUCKER, W. D. (1988) Plasma immunoreactive propiomelanocortin peptides and cortisol in normal dogs and dogs with Cushing's syndrome: diurnal rhythm and response to various stimuli. *Endocrinology*, **122**, 1250–62.

PLANT, T. M. (1981) Time courses of concentrations of circulating gonadotrophin, prolactin, testosterone and cortisol in adult male rhesus monkeys (*Macaca mulatta*) throughout the 24 h light–dark cycle. *Biology and Reproduction*, **25**, 244–52.

POPPER, C. W., CHIEUH, C. C. & KOPIN, J. J. (1978) Plasma catecholamine concentrations in unanaesthetized rats during sleep, wakefulness, immobilization and after decapitation. *Journal of Pharmacology and Experimental Therapeutics*, **202**, 144–8.

RATCLIFFE, J. M., MCELHATTON, P. R. & SULLIVAN, F. M. (1993) Reproductive toxicity. In Ballantyne, B., Marrs, T. & Turner, P. (eds), *General and Applied Toxicology*, pp. 989–1021. New York: Stockton Press.

READER, S. C. J., SHINGLES, C. & STONARD, M. D. (1991) Acute testicular toxicity of 1:3 dinitro benzene and ethylene glycol monomethyl ether in the rat: evaluation of biochemical effect markers and hormonal responses. *Fundamental and Applied Toxicology*, **16**, 61–70.

REICHLIN, S. (1985) Neuroendocrinology. In Wilson, J. D. & Foster, D. W. (eds), *Williams Textbook of Endocrinology*, 7th Edn, pp. 492–567. Philadelphia: W. B. Saunders.

RIBELIN, W. E. (1984) Effects of drugs and chemicals upon the structure of the adrenal gland. *Fundamental and Applied Toxicology*, **4**, 105–19.

SANTEN, R. J. & BARDIN, C. W. (1973) Episodic LH secretion in man: pulse analysis, clinical interpretation, physiological mechanisms. *Journal of Clinical Investigation*, **52**, 2617–28.

SEAL, U. S. & DOE, R. P. (1966) Vertebrate distribution of corticosteroid-binding globulin and some endocrine effects on concentration. *Steroids*, **5**, 827–41.

SELBY, P. P. & ADAMS, P. H. (1994) The investigation of hypercalcaemia. *Journal of Clinical Pathology*, **47**, 579–84.

SHARPE, R. M. (1982) The hormonal regulation of the Leydig cell. In Finn, C. A. (ed.), *Reviews of Reproductive Biology*, Vol. 4, pp. 241–317. Oxford: Clarendon Press.

SPITZ, I. M., GUNSALUS, G. L., MATHER, J. P., THAU, R. & BARDIN, C. W. (1985) The effects of imidazole carboxylic acid derivative, tolmidamine on testicular function. 1. Early changes in androgen binding protein secretion in the rat. *Journal of Androgens*, **6**, 161–78.

SZABO, S. & LIPPE, I. T. (1989) Adrenal gland: chemically induced structural and functional changes in the cortex. *Toxicologic Pathology*, **17**, 317–29.

TALMAGE, R. V., DOPPELT, S. H. & COOPER, C. W. (1975) Relationship of blood concentrations of calcium, phosphate, gastrin and calcitonin to the onset of feeding in the rat. *Proceedings of the Society for Experimental Biology and Medicine*, **149**, 855–9.

THOMAS, J. A. & KEENAN, E. G. (1986) *Principles of Pharmacology*, pp. 1–294. New York: Plenum Press.

TIMBRELL, J. A., DRAPER, R. & WATERFIELD, C. J. (1994) Biomarkers in Toxicology. New uses for some old molecules. *Toxicology and Ecotoxicology News*, **1**, 4–14.

VERNAY, M., MARTY, J. & MOALTI, J. (1984) Absorption of electrolytes and volatile fatty acid in the hind-gut of the rabbit. Circadian rhythm of hind-gut electrolytes and plasma aldosterone. *British Journal of Nutrition*, **52**, 419–28.

WESTPHAL, U. (1971) *Steroid Protein Interactions*, pp. 164–350. Berlin: Springer-Verlag.

WHEAT, T. E., HINTZ, M., GOLDBERG, E. & MARGOLIASH, E. (1977) Analyses of stage specific multiple forms of lactate dehydrogenase and of cytochrome c during spermatogenesis in the mouse. *Differentiation*, **9**, 37–41.

WILSON, C. A. & LEIGH, A. J. (1992) Endocrine toxicology of the female reproductive system. In Atterwill, C. K. and Flack, J. (eds), *Endocrine Toxicology*, pp. 313–99. Cambridge: Cambridge University Press.

WONG, C.-C., DOHLER, K.-D., GEAELINGS, H. & VON ZUR MUHLEN, A. (1983) Influence of age, strain and season on circadian periodicity of pituitary, gonadal and adrenal hormones in the serum of male laboratory rats. *Hormone Research*, **17**, 202–15.

9

Assessment of Gastrointestinal Toxicity and Pancreatic Toxicity

G. O. EVANS

The disposition of xenobiotics is often considered to occur in four interrelated phases, i.e. absorption, distribution, metabolism and excretion (collectively abbreviated to ADME). The major routes for the absorption of xenobiotics include skin, lungs (or gills), the parenteral routes – intravenous (i.v.), intraperitoneal (i.p.), subcutaneous (s.c.) and intramuscular (i.m.) – but the commonest route for administration in toxicology studies is per orally (p.o.), with subsequent absorption via the gastrointestinal (GI) tract.

In the GI tract, xenobiotics may be hydrolysed by gastric acidity, metabolized by intestinal bacteria, or the test compound (and/or its metabolites) may be excreted via the bile following passage through the liver. Some xenobiotics (and/or metabolites) which undergo biliary excretion are then reabsorbed in the intestine, i.e. entero-hepatic cycling.

The potential for the absorption of xenobiotics (or metabolites) is governed by molecular size and charge, hydrophobic properties (or lipid solubility), and the structure and similarities to endogenous molecules. Mechanisms exist for both active and passive transport of test compounds via the GI tract (Timbrell, 1994). Simple diffusion processes permit the absorption of non-ionic hydrophilic substances, e.g. urea, and anionic substances with hydration radii of less than 2.9 Å, e.g. nitrates. Many lipophilic compounds such as barbiturates, food additives and industrial solvents are absorbed by passive diffusion. The term 'facilitated diffusion process' is used to describe some absorption processes which act down a concentration gradient involving a carrier protein, but do not have a requirement for metabolic energy. Cations are often absorbed by a carrier-mediated transport system which is energy dependent.

Given these differing transport mechanisms and the diverse tissues of the GI tract, it is not surprising that anatomical differences between species lead to variations of physiological, ADME and toxicological effects, although severely irritant and corrosive materials which produce toxic effects by direct action on the walls of the GI tract often produce these effects in several species.

Nutritional status may influence the bioavailability and hence toxic effects on the gastrointestinal tract in several ways (George, 1984; Omaye, 1985; Aungst and

Shen, 1986). In animal toxicological studies, there is an opportunity to obtain data on food and fluid intake during the treatment period, and then to use these data (for the different treatment groups and controls) as an indicator of gastrointestinal function. In many studies of foods and food additives, much of the literature is focused on effects on malignant and vascular disease states, with relatively few clinical chemistry measurements.

It is generally thought that rodents and rabbits do not vomit while dogs, non-human primates and cats do exhibit this reflex: when stimulated, frequent emesis can affect the absorption, toxicity and several biochemical measurements. Using oral gavage, where the gavage tube is correctly sited in the stomach, generally avoids injury to the oesophagus, particularly in rodents where regurgitation into the oesophagus does not occur. It is well recognized that exposure by inhalation may result in secondary exposure via the GI tract.

While aqueous vehicles such as water or saline may be preferred, some non-soluble materials may be administered with other vehicles such as oils or organic solvents which may exert separate effects. Ingestion of significant quantities of corn oil can increase the flow of lymph and affect uptake of lipophilic compounds.

In general toxicology studies, the selection of clinical chemistry tests is generally focused on fluid and electrolyte balance supported by enzyme, carbohydrate, lipid and protein measurements, but few of the common tests for regulatory studies are specific for gastrointestinal toxicity (see Chapter 2).

9.1 Electrolyte and Fluid Balance (see Chapter 14)

Disturbances of gastrointestinal function are often accompanied by electrolyte imbalance due to fluid losses either by emesis (vomiting), volvulus (dilation), diarrhoea or other perturbations of many varied mechanisms for electrolyte and fluid homeostasis. Excessive and prolonged salivation may also cause electrolyte perturbations. Prolonged or excessive losses of fluid via the GI tract will affect packed cell volume (haematocrit), plasma total protein, albumin, electrolytes, acid–base balance and osmolality values (Smith, 1986).

Hypo- or hyper-natraemia may occur depending on the proportional losses of electrolyte to water; these electrolyte changes are also reflected by plasma osmolality. In the presence of hyperlipidaemia and hyperproteinaemia there may be significant differences between measured and calculated plasma osmolality.

The pancreatic secretion of chloride varies inversely to the biocarbonate concentration which in turn varies directly with the flow rate: the sum of these two anions tends to remain constant under normal circumstances, with the electrolyte concentrations of the pancreatic secretions tending to parallel blood pH and electrolyte concentrations. Excessive losses of pancreatic fluid can be monitored by plasma chloride measurements: changes of plasma chloride concentration which are not accompanied by a change in plasma sodium are usually associated with disturbances of acid–base balance. Hypocalcaemia may also accompany severe pancreatic toxicity: this is probably due to the formation of insoluble salts of fatty acids.

9.2 Carbohydrate and Lipid Metabolism

Body glucose is mainly confined to the extracellular fluid with small freely exchangeable amounts in the liver and erythrocytes. Glucose enters the body pool from either nutritional sources or by hepatic glycogenolysis. Although this glucose pool is generally constant in laboratory animals, it varies more than in man. The liver plays a key role in maintaining glucose homeostasis removing approximately 70% of the glucose load via the portal circulation, and storing this as glycogen. In addition, the precursors – lactate, glycerol, pyruvate and alanine – resulting from tissue metabolism are converted into glucose by hepatic gluconeogenesis. Thus, it is predictable that hepatotoxicity may be accompanied by perturbations of carbohydrate metabolism. Excessive lipid catabolism associated with carbohydrate perturbations may lead to ketonuria. The presence of glucosuria in the absence of hyperglycaemia may indicate renal tubular injury (see Chapter 7).

Several hormones are concerned with regulating glucose metabolism in the fasting and non-fasting states, and these hormones include insulin, glucagon, growth hormone, adrenaline and cortisol. Although insulin is the main hormone, playing an important role in the synthesis of proteins, triglycerides and polysaccharides, the levels of this hormone change rapidly and data obtained from toxicology studies may be highly variable.

Plasma and urinary glucose together with additional measurements for ketosis, i.e. urinary ketones, plasma 3-hydroxybutyrate, lipids (Chapter 13) and osmolality measurements can be used to monitor the severity of disturbed carbohydrate metabolism. Electrolyte measurements, particularly sodium and potassium, may be altered in severe disturbances of carbohydrate metabolism. The additional measurement of glycosylated haemoglobin may be useful when monitoring the effects of hypoglycaemic agents (Higgins *et al.*, 1982; Srinivas *et al.*, 1986). Although glucose measurements may be of particular pharmacological interest when testing xenobiotics designed as hypo- or hyperglycaemic agents, it is rarely necessary to use glucose tolerance tests as part of toxicological studies.

9.2.1 *Plasma or Serum Glucose*

The use of an inhibitor, i.e. sodium fluoride/oxalate, or rapid separation of the plasma is recommended to prevent glycolysis due to the presence of erythrocytic enzymes. Blood glucose values measured with a dry film/test strip system may be erroneous where the haematocrit values are outside the reference ranges. The ability of certain primates to store food in their buccal pouches often prevents the true measurement of 'fasting' glucose. Plasma glucose levels below 1.7 mmol/L, which would give rise to clinical signs of hypoglycaemia in man, do not appear to produce similar adverse effects in some other primate species. With some rodent species, it appears that the animals have to be fasted for much longer periods than other species to achieve similar reductions of plasma glucose. This latter observation has led to some debate about the merits of fasting smaller laboratory animals such as mice and rats prior to blood collection (see Chapter 2). Hypoglycaemia is apparent in some animals where gastric absorption or food intake has been markedly altered. Blood collection procedures may cause a marked elevation of plasma glucose where the animal is subject to stress including restraint (Gartner *et al.*, 1980).

9.2.2 *Urine Glucose*

Test strips (or dipsticks) commonly used for urinary glucose are impregnated with glucose oxidase, and these yield semi-quantitative results; these test strips can be affected by urinary ascorbate (see Chapter 3). Semi-quantitative methods, using a tablet form of alkaline copper sulphate reagent, can be used to measure glycosuria; these tests will detect reducing substances other than glucose, e.g. fructose. More laboratories are turning to quantitative methods particularly for the detection of renal tubular injury.

An increased production of acetoacetate, 3-hydroxybutyrate and acetone collectively termed 'ketone bodies' is associated with hyperglycaemia, and this may be detected as ketonuria.

9.3 Enzymes

Several enzyme measurements are available for assessing gastrointestinal function and toxicity, but these assays (apart from amylase and possibly lipase) are not usually included in the majority of toxicology studies. The pancreas secretes more than 20 different inactive precursors of serine proteases and exopeptidases, lipase and amylase. Of these enzymes, the major proteolytic enzymes (and their inactive proenzymes) are trypsin (trypsinogen), chymotrypsin (chymotrypsinogen) and carboxypeptidases (procarboxypeptidase). Relatively simple assays are used for amylase and lipase (Boyd *et al.*, 1988), whereas the measurement of other enzymes, e.g. intestinal disaccharidases, is more complex: these are subject to various factors, including age, nutritional status, hormonal influences due to glucocorticoids and thyroid hormones, diurnal variations etc. (Henning, 1984). Some enzyme measurements require the invasive collection of gastric, pancreatic, intestinal fluids or tissues and therefore are not suitable for many toxicological studies.

9.3.1 *Amylase*

Apart from pancreatic tissue, amylase is found in other tissues such as the liver, salivary glands and small intestine. There are wide interspecies differences for the plasma amylase levels, and tissue distribution of amylase (Rajasingham *et al.*, 1971; McGeachin and Akin, 1982). The number of isoenzymes in different species varies, and plasma amylase in the plasma may be the pancreatic (P) isoamylase or (S) isoamylase from the salivary gland: in mice it is reported that serum amylase is of salivary origin while urinary amylase is of pancreatic origin (MacKenzie and Mosser, 1976). Total plasma amylase is more commonly measured as salivary changes are relatively uncommon, but they do occur in some conditions such as sialoadenosis (Arglebe *et al.*, 1978). The plasma P amylase is probably a more sensitive and specific assay for pancreatitis in some species where this assay works.

The half-life of amylase is short, varying from less than 2 h in the mouse to 5 h in the dog. Urinary amylase is generally increased where plasma amylase is increased, but it offers no major diagnostic advantage over plasma measurements because it is also affected by renal injury and disease.

9.3.2 Lipase

Pancreatic lipase is secreted in its active form and this activity is enhanced by colipase and bile salts. Other lipases – phospholipase a, phospholipase b and cholesterol ester hydrolase – are also secreted by the pancreas. Following the introduction of simpler assays, plasma lipase measurements are being used increasingly and their use in conjunction with plasma amylase can help diagnosis as several drugs can affect the pancreas (Banerjee *et al.*, 1989). Some drugs, such as dexamethasone in dogs, appear to alter serum lipase more than serum amylase (Parent, 1982).

9.3.3 Proteolytic Enzymes

Pepsinogen is the precursor of the proteolytic enzyme pepsin and is secreted by gastric parietal cells; it may be measured in plasma or gastric fluid using colorimetric, fluorimetric or radioimmunometric methods (Will *et al.*, 1984; Ford *et al.*, 1985; Tani *et al.*, 1987). Pepsinogen activities may be increased following peptic ulceration and by parasitic infections.

9.3.4 Alkaline Phosphatase

Another plasma enzyme of interest in rodent studies is alkaline phosphatase which has a much larger intestinal isoenzyme component than in many other species (Unakami *et al.*, 1989; see Chapter 5). Duodenal or intestinal alkaline phosphatase is markedly reduced with some ulcerogens, e.g. cysteamine (Japundzic and Levi, 1987).

Enzyme changes occur in several gastrointestinal conditions such as intestinal infarction or obstruction (Kazmlerczak *et al.*, 1988), parasitic infections, obstruction of the biliary system (Yoshida and Koga, 1988), contraction of the sphincter of Oddi by drugs such as morphine (Webster and Zieve, 1962), withdrawal of food (Kuhn and Hardegg, 1988), and age-related changes (Denda *et al.*, 1994).

9.4 Hormones

The gut hormones (Table 9.1) are generally of low molecular mass, and many are phylogenetically and structurally related. The number of constituent amino acids ranges from 14 for somatostatin to 43 for GIP; insulin is composed of two peptide chains. The differences of structure between species for insulin and glucagon prevent the universal application of immunoassays across different species, although there are degrees of cross-reactivity between some species (Young, 1963;

Table 9.1 Examples of pancreatic and gut hormones

Hormone and location	Function
Gastric antrum and duodenum	
Gastrin (from G cells)	Stimulates gastric H^+ secretion and trophic to gastric mucosa
Duodenum and jejunum	
Secretin	Stimulates pancreatic secretion of bicarbonate and water
Cholecystokinin (CCK)	Stimulates secretion of pancreatic enzymes
Glucose-dependent insulinotrophic peptide (GIP)	Stimulates postprandial release of insulin, inhibits gastric motility
Motilin	Stimulates intestinal motor activity
Pancreas	
Insulin	Stimulates glycogen synthesis, glucose uptake and inhibits lipolysis
Glucagon	Opposite actions to insulin
Somatostatin	Reduces secretion of insulin CCK, GIP and VIP
Pancreatic polypeptide	Inhibits release of pancreatic enzymes (and relaxes gallbladder where present)
Ileum and colon	
Enteroglucagon	Increases small intestinal mucosal growth and retards intestinal transit rate
All areas of GI tract	
Vasoactive intestinal polypeptide (VIP)	Affects secretomotor actions, vasodilation, relaxation of intestinal smooth muscle

Berthet, 1963). Reactivities with antisera to insulin from dogs, Old-World monkeys and humans are generally equivalent but some species such as the guinea pig show much lower activities. In general gut hormone measurements are reserved for mechanistic studies.

Plasma gastrin measurements may be a useful adjunct to pharmacological studies where anti-ulceration drugs are being investigated, but some of the problems associated with this hormone illustrate the general difficulties of measuring gut hormones. Given this hormone is secreted in response to distension of the stomach, the presence of some amino acids and peptides following protein digestion, or fall in gastric pH in the presence of highly alkaline food or xenobiotics, plasma levels can vary considerably and therefore should be measured in fasting animals. Plasma gastrin has a relatively short half-life (Koop *et al.*, 1982; Larsson *et al.*, 1986), and the peptide is labile requiring special sample collection procedures with protease inhibitors.

9.5 Other Tests of Gastrointestinal and Pancreatic Function

Pancreatic function tests include the use of synthetic peptide N-benzoyl-1-tyrosyl-4-aminobenzoic acid (BT-PABA) tests, fluoroscein dilaurate, disaccharide tolerance, fat absorption, and stimulation of pancreatic enzyme secretion by cholecystokinin-pancreozymin (CCK-PZ) or the Lundh test meal. As the exocrine pancreas is a target for toxicity from oxygen free radicals, other measurements to assess the effect of these free radicals may be used in mechanistic studies (Braganza, 1993). The use of functional stimulation tests such as pentagastrin and xylose absorption tests are relatively uncommon in toxicological studies, although these tests are used in veterinary medicine. Where gastrointestinal toxic effects are prolonged, marked reductions in the intake of essential nutrients such as vitamins (folate), iron, amino acids etc. may occur and this may lead to additional measurements.

9.6 Faecal Tests

9.6.1 *Occult Blood*

This assay can be used to assess the integrity of the mucosal lining of the GI tract where gastrointestinal bleeding is suspected but blood cannot be seen macroscopically in the faeces, e.g. with non-steroidal anti-inflammatory compounds (Szabo *et al.*, 1989; Walsh, 1989). The sampling procedures and their timing are important particularly where the bleeding may be intermittent.

There are a number of available procedures for the detection of occult blood, and the majority of colorimetric qualitative tests are based on the pseudo-peroxidase activity of haemoglobin. The sensitivity of these tests varies with the reagents used, and in animal studies it may be necessary to alter the sensitivity of a particular method in order to avoid false positive reactions. Additionally, false positive reactions may be caused by some cleaning fluids used in animal housing, e.g. hypochlorite solutions, and false negative effects also may occur with some compounds, e.g. ascorbate.

Alternative methods for the detection of faecal occult blood include the use of erythrocytes labelled with chromium or iron radioisotopes (Walsh, 1989), or the detection of porphyrins by fluorescence (Boulay *et al.*, 1986), but these methods are not used widely.

9.6.2 *Fat*

The observation of the presence of gross faecal fat content can indicate effects on pancreatic function or biliary dysfunction. In longer-term studies, malabsorption of fat-soluble vitamins may be reflected by the clinical condition of vitamin-deficient animals.

References

Arglebe, C., Bremer, K. & Chilla, R. (1978) Hyperamylasaemia in isoprenaline-induced experimental sialoadenosis in the rat. *Archives of Oral Biology*, **23**, 997–9.

Aungst, B. & Shen, D. D. (1986) In Rozman, K. & Hanninen, O. (eds), *Gastrointestinal Toxicology*, pp. 29–55. Amsterdam: Elsevier Science.

Banerjee, A. K., Patel, K. J. & Grainger, S. L. (1989) Drug-induced acute pancreatitis. A critical review. *Medical Toxicology and Adverse Drug Experience*, **4**, 146–98.

BERTHET, J. (1963). Pancreatic hormones: glucagon. In von Euler, U. S. & Heller, H. (eds), *Comparative Endocrinology*, pp. 410–28. London: Academic Press.

BOULAY, J. P., LIPOWITZ, A. J., KLAUSNER, J. S., ELLEFSON, M. L. & SCHWARTZ, S. (1986) Evaluation of a fluorimetric method for the quantitative assay of fecal hemoglobin in the dog. *American Journal of Veterinary Research*, **47**, 1293–5.

BOYD, E. J. S., RINDERKNECHT, H. & WORMSLEY, K. G. (1988) Laboratory tests in the diagnosis of the chronic pancreatic diseases. Part 4. Tests involving the measurement of pancreatic enzymes in body fluid. *International Journal of Pancreatology*, **3**, 1–16.

BRAGANZA, J. M. (1993) Toxicology of the pancreas. In Ballantyne, B., Marrs, T. & Turner, P. (eds), *General Toxicology*, pp. 663–715. New York: Stockton Press.

DENDA, A., TSUTSUMI, M. & KONISHI, Y. (1994) Exocrine pancreas. In Mohr, U., Dungworth, D. L. & Capen, C. C. (eds), *Pathobiology of the Aging Rat*, Vol. 2, pp. 351–61. Washington: International Life Sciences Institute.

FORD, T. F., GRANT, D. A. W., AUSTEN, B. M. & HERMON-TAYLOR, J. (1985) Intramucosal activation of pepsinogens in the pathogenesis of acute gastric erosions and their prevention by the potent semisynthetic amphipathic inhibitor pepstatinyl-glycyl-lysyl-lysine. *Clinica Chimica Acta*, **145**, 37–47.

GARTNER, K., BUTTNER, D., DOHLER, K., FRIEDEL, R., LINDENA, J. & TRAUTSCHOLD, I. (1980) Stress response of rats to handling and experimental procedures. *Laboratory Animals*, **14**, 267–74.

GEORGE, C. F. (1984) Food, drugs, and bioavailability. *British Medical Journal*, **289**, 1093–4.

HENNING, S. J. (1984) Hormonal and dietary regulation of intestinal enzyme development. In Schiller, C. M. (ed.), *Intestinal Toxicology*. New York: Raven Press.

HIGGINS, P. J., GARLICK, R. L. & BUNN, H. F. (1982) Glycolsylated hemoglobin in human and animal red cells. *Diabetes*, **26**, 743–8.

JAPUNDZIC, I. & LEVI, E. (1987) Mechanism of action of cysteamine on duodenal alkaline phosphatase. *Biochemical Pharmacology*, **36**, 2489–95.

KAZMLERCZAK, S. C., LOTT, J. A. & CALDWELL, J. H. (1988) Acute intestinal infarction or obstruction: search for better laboratory tests in an animal model. *Clinical Chemistry*, **34**, 281–8.

KOOP, H., SCHWAB, E., ARNOLD, R. & CREUTZFELDT, W. (1982) Effect of food deprivation on rat gastric somatostatin and gastrin release. *Gastroenterology*, **82**, 871–6.

KUHN, G. & HARDEGG, W. (1988) Plasma amylase and lipase activities in dogs with variation in food composition and availability. In Beynen, A. C. & Solleveld, H. A. (eds), *New Developments in Biosciences: Their Implications for Laboratory Animal Science*, pp. 365–71. Dordrecht: Martinus Nijhoff.

LARSSON, H., CARLSSON, E., MATTSON, H., LUNDELL, L., SUNDLER, F., SUNDELL, G., WALLMARK, B., WANATABE, T. & HAKANSON, R. (1986) Plasma gastrin and gastric enterochromaffinlike cell activation and proliferation. *Gastroenterology*, **90**, 391–9.

MACKENZIE, P. I. & MOSSER, M. (1976) Studies on the origin and excretion of serum α amylase in the mouse. *Comparative Biochemistry and Physiology*, **54B**, 103–6.

MCGEACHIN, R. L. & AKIN, J. R. (1982) Amylase levels in the tissues and body fluids of several primate species. *Comparative Biochemistry and Physiology*, **72A**, 267–9.

OMAYE, S. T. (1985) Effects of diet on toxicity testing. *Federation Proceedings*, **45**, 133–5.

PARENT, J. (1982) Effects of dexamethasone on pancreatic tissue and on serum amylase and lipase activities in dogs. *Journal of the Veterinary Medical Association*, **180**, 743–6.

RAJASINGHAM, R., BELL, J. L. & BARON, D. N. (1971) A comparative study of the isoenzymes of mammalian alpha amylase. *Enzyme*, **12**, 180–6.

SMITH, P. L. (1986) Gastrointestinal physiology. In Rozman, K. and Hanninen, O. (eds), *Gastrointestinal Toxicology*. Amsterdam: Elsevier Science.

SRINIVAS, M., GHOSH, K., SHOME, D. K., VIRDI, J. S., KUMAR, S., MOHANTY, D. & DAS, K. C. (1986) Glycosylated hemoglobin (HbA_1) in normal rhesus monkeys (*Macaca mulatta*). *Journal of Medical Primatology*, **15**, 361–5.

SZABO, S., SPILL, W. F. & RAINSFORD, K. D. (1989) Non-steroidal anti-inflammatory drug-induced gastropathy. Mechanism and management. *Medical Toxicology and Adverse Drug Experience*, **4**, 77–94.

TANI, S., ISHIKAWA, A., YAMAZAKI, H. & KUDO, Y. (1987) Serum pepsinogen levels in normal and experimental peptic ulcer rats measured by radioimmunoassay. *Chemical and Pharmacology Bulletin*, **35**, 1515–22.

TIMBRELL, J. (1994) *Principles of Biochemical Toxicology*, 2nd Edn. London: Taylor & Francis.

UNAKAMI, S., HIRATA, M., ICHINOHE, K., TANIMOTO, Y. & HZUKA, H. (1989) Separation of and quantification of serum alkaline phosphatase isozymes in the rat by affinity electrophoresis. *Experimental Animals*, **38**, 85–9.

WALSH, C. T. (1989) Methods in gastrointestinal toxicology. In Hayes, A. W. (ed.), *Principles and Methods of Toxicology*, 2nd Edn. pp. 659–72. New York: Raven Press.

WEBSTER, P. D. & ZIEVE, L. (1962) Alterations in serum content of pancreatic enzymes. *New England Journal of Medicine*, **267**, 604–7.

WILL, P. C., ALLBEE, W. E., WITT, C. G., BERTKO, R. J. & GAGINELLA, T. S. (1984) Quantification of pepsin A activity in canine and rat gastric juice with the chromogenic substrate Azocoll. *Clinical Chemistry*, **30**, 707–11.

YOSHIDA, M. & KOGA, A. (1988) Alterations with time of secretion and ultrastructure of the exocrine pancreas in rats with an obstructed biliary system. *British Journal of Experimental Pathology*, **69**, 441–8.

YOUNG, F. G. (1963) Pancreatic hormones: insulin. In von Euler, U. S. and Heller, H. (eds), *Comparative Endocrinology*, pp. 371–410. London: Academic Press.

10

Assessment of Cardiotoxicity and Myotoxicity

G. O. EVANS

10.1 Cardiotoxicity

Cardiac damage can be caused by many xenobiotics, and the effects may be acute or chronic depending on the exposure time and action of the compound. Chemicals can selectively affect the heart or vasculature causing inflammatory and degenerative changes, which may then lead to impairment of the circulation. Primary cardiovascular toxicity is most commonly associated with exaggerated pharmacological effects at high dosages: mechanistically, these changes may involve perturbations of membrane function, particularly of ion transport, and in the contractile or energy-generating systems of the tissues.

Adverse secondary cardiovascular changes may follow toxic effects on other organs, e.g. kidney, liver, thyroid and adrenal cortex, or be the result of pharmacological effects on non-cardiac tissue, e.g. drugs affecting the central nervous system. Some cardiotoxins act as haptens, binding directly with proteins or nucleic acids, or they may act indirectly on the immune system. These immune reactions generally occur in small blood vessels and can involve the deposition of immune complexes.

The choice of an appropriate animal species may vary with the pharmacological action of the test compound (Czarnecki, 1984), and occasionally animal models of cardiac disease may be appropriate for the study of paradoxical effects, such as those observed with some bronchodilators. Histopathology and electrocardiography remain the primary methods for the assessment of functional and degenerative effects on cardiac tissue. Various biochemical tests have been suggested for monitoring laboratory animals during toxicology studies, but none of these tests offers absolute specificity for the detection of cardiovascular changes. Provided that the functional interrelationships between the heart and other organs, particularly those of the liver and kidneys, are considered, then some tests may be of value. Obesity, diet, genetic variation, anaemia, thyroid status, and poor nutritional status may all be contributing factors to cardiotoxicity.

Injury to cardiac and skeletal muscle can result in the leakage of cellular contents into the circulating blood system with early loss of ions and metabolites.

Table 10.1 Plasma enzymes and their respective half-lives (h) in two species (after Lindena *et al.*, 1986)

Enzyme	Dog	Rat
AST	3.3 to 4.4	2.3
CK (total)	0.6 to 16.2	0.6
CK-MB	1.3 to 8.1	–
LDH (total)	1.6	–
LDH - H4	3.3	5.3
LDH - M4	0.5 to 1.8	0.4 to 0.9

The release of macromolecules, including enzymes and proteins following changes of cellular permeability and necrosis, has been the primary focus for plasma measurements in cardiotoxicity. Although haemoglobinuria, proteinuria and enzymuria may occur following myocardial infarction or damage to pulmonary arteries, urine tests are generally not helpful in the detection of cardiotoxicity.

10.1.1 Enzymes

Following cardiac damage, a large number of tissue enzymes are released into the plasma and these can be used as markers of cardiotoxicity. However, the timing and the methods for sample collection are particularly critical, as the half-lives of the enzymes differ (Table 10.1). The enzymes creatine kinase (CK), lactate dehydrogenase (LDH), aspartate aminotransferase (AST) and to a lesser extent alanine aminotransferase (ALT) have been used in cardiotoxicity studies: none of these enzyme measurements is specific for cardiac tissue. Additionally, the tissue distribution of these enzymes varies with the species (see Chapter 5 – General Enzymology). These enzyme measurements are of particular value in the assessment of thrombolytic agents in animal models, where serial measurements are made in acute studies. This sampling pattern and successful application of enzyme measurements in cardiac studies contrast with the problems of performing enzyme measurements at longer time intervals in regulatory toxicology studies, where the enzymes may be measured several weeks after tissue damage has occurred.

Creatine kinase (CK) is a dimeric molecule with subunits B and M; there are three cytosolic enzymes, the 'muscle type' dimer (CK-MM), the 'brain type' dimer (CK-BB) and the 'myocardial type' dimer (CK-MB) (Lang, 1981). Mitochondrial isoenzymes and isoforms of CK also exist (Braun, 1992). More recently it has been recognized that these isoenzymes are glycoproteins (McBride *et al.*, 1990). The isoenzyme distribution varies between tissues; for example, in the baboon, the dominant isoenzyme in the myocardium is CK-MM while CK-BB is the dominant isoenzyme in the tissues of the gastrointestinal tract (Yasmineh *et al.*, 1976).

Lactate dehydrogenase (LDH) is a cytosolic tetrameric enzyme with five major isoenzymes in plasma consisting of H (heart) and M (muscle) subunits. The five isoenzymes are numbered according to decreasing anodic mobility during electrophoretic separation: LDH_1 has four H subunits, LDH_5 has four M subunits, and LDH_2, LDH_3 and LDH_4 are hybrid combinations – containing HHHM,

HHMM and HMMM respectively (Markert and Whitt, 1975). The widespread tissue distribution of LDH differs with the various species (Garbus *et al.*, 1967; Karlsson and Larsson, 1971; Schultze *et al.*, 1994). Additional isoenzymes of LDH have been described, and some LDH isoenzymes may complex with drugs, e.g. streptokinase (Poldasek and McPherson, 1989).

Preanalytical factors can alter plasma or serum levels of both total CK and LDH. Some of these factors are mentioned in Chapter 3, and have been reviewed for rats elsewhere (Evans, 1991; Matsuzawa and Ishikawa, 1993). Other factors to be considered include circadian rhythms in rats (Maejima and Nagase, 1991), incorrect venepuncture in dogs (Fayolle *et al.*, 1992), and macrophage clearance of LDH_5 in mice (Hayashi and Notkins, 1994).

The broad normal (or reference) ranges for plasma total LDH activity encountered in laboratory animals make interpretation difficult (Matsuzawa *et al.*, 1993). In the rat, LDH_5 is a major isoenzyme in plasma while LDH_2 is the major isoenzyme of heart tissue (Garbus *et al.*, 1967; Dimitrijevic *et al.*, 1993); therefore a considerable release of LDH_2 from cardiac tissue must occur before total plasma LDH values change significantly. Plasma alpha-hydroxybutyrate dehydrogenase (HBD) measurements can reflect the activities of LDH_1 and LDH_2 isoenzymes in the rat if the appropriate analytical conditions are employed (Barrett *et al.*, 1988).

The relative insensitivity of total plasma CK and LDH measurements has led to the use of isoenzyme measurements. CK and LDH isoenzymes can be separated by a variety of electrophoretic techniques or immuno-inhibition methods, but the latter are limited by the specificity of the monoclonal antibodies currently available (Lang, 1981; Landt *et al.*, 1989). Some confusion exists in published data concerning the plasma CK isoenzyme patterns in relatively common laboratory animals such as the dog and the rat (Evans, 1991), and investigators are advised to establish reference patterns for control animals.

Several investigators have reported enzyme changes using a classical model of cardiotoxicity induced by isoprenaline: some of these enzyme changes are the result of myocardial hypoxia in addition to cellular necrosis (Meltzer and Guschwan, 1971; Sadek and Pfitzer, 1975; Balazs and Bloom, 1982; Barrett *et al.*, 1988; Bhargava *et al.*, 1990). These data also illustrate the careful attention to the timing of sample collections.

Another model of cardiotoxicity is that induced by the anthracyclines such as adriamycin, where enzyme changes have been described (Olson and Capen, 1977). However, Zbinden and colleagues (1991) reported that plasma enzymes were unreliable indicators of myocardial damage caused by these compounds. Transient elevations of serum CK and LDH suggestive of myocardial damage may occur in the absence of histopathological evidence of cardiotoxicity, e.g. with amsacrine (Kim *et al.*, 1985).

10.1.2 Electrolytes

Disturbances of the intra- and extracellular equilibria of the cations – sodium, potassium, calcium and magnesium – may result in increased irritability of cardiac tissue and be associated with arrhythmias. The balance of these cations is interrelated, so a severe alteration of one cation may alter the other cations. As the divalent cations – calcium and magnesium – are partially bound to plasma proteins, particularly albumin, it is necessary to consider the metabolically more important ionized fractions in plasma, and to recognize that this cationic binding varies with animal species. Electrolyte perturbations are observed with compounds such as digitalis, glycosides, mineralocorticoids and lithium. Iron chelation by doxorubicin is associated with changes of free radicals and subsequent cellular damage (Myers, 1988).

Cationic changes are accompanied invariably by disturbances of plasma anionic concentrations and acid–base balance. Where there are alterations of fluid balance such as occur with congestive cardiac failure, these may be reflected by changes in plasma osmolality and proteinic concentrations, particularly albumin. Plasma anions alone are generally poor markers of cardiotoxicity as these may alter in many different conditions. Plasma osmolality measurements can be useful where major changes in body fluids occur: when large amounts of plasma are lost during inflammatory processes by haemorrhage, or if fluid loss occurs via the gastrointestinal tract or vomitus. Alterations of plasma electrolytes and osmolality can occur as a consequence of circulatory insufficiency, and the interpretation of changing values for plasma electrolytes is complicated where changes of cardiac output occur as a consequence of the failure of other organ functions, e.g. the kidneys, and vice versa. Further measurements for electrolyte homeostasis may include components of the renin–angiotensin axis and atrial natriuretic factor (Laragh, 1985; see Chapter 12).

10.1.3 Proteins

Following cardiotoxic changes, plasma protein patterns may alter, with increases of acute phase proteins, e.g. myoglobin, C-reactive protein, fibrinogen and other related coagulation proteins. These changes may be detected by electrophoretic techniques or selective methods for quantifying individual protein fractions (see Chapter 14).

10.1.4 Lipids

The roles of lipids in cardiotoxicity are discussed in Chapter 13. Plasma lipids are indicators of potential risks for cardiotoxicity contrasting with some of the other markers discussed here, which directly or indirectly reflect damage to cardiac tissues.

10.1.5 Troponins

There has been recent interest in assays for troponins, which are the protein filament components of the contractile cardiac and skeletal muscles, but which are not present in smooth muscle. There are three genetically and biochemically distinct troponins: these are troponin T (TnT) which binds to tropomyosin, troponin I (TnI) which inhibits myosin ATPase and troponin C (TnC) which binds to calcium (Katus *et al.*, 1992). At present, it is unclear if immunoassays can be developed to differentiate between skeletal and cardiac troponins in laboratory animals, and which cardiac troponin is the better marker of cardiac damage (Saggin *et al.*, 1990; Malouf *et al.*, 1992). If the cardiac troponins persist longer than CK (and its isoenzymes) in the plasma, then these assays may be suitable, but further evaluations are required before their widespread use in toxicology studies. Voss and colleagues (1994) reported the use of cardiac TnT in dogs following coronary artery occlusion; they found serum CK-MB and cardiac TnT profiles were similar to that observed in humans, but with a poor correlation to infarct size.

10.2 Myotoxicity

The mechanisms which cause myotoxicity of skeletal muscle include disruption of the normal membrane function and changes in energy associated with the mitochondria. The skeletal muscle fibres are not homogenous and there are at least two major types – the type I or the slow twitch and the fast twitch type II – which can be distinguished by histological techniques. Using the common plasma enzyme measurements, it is impossible to distinguish between cardiotoxicity and myotoxicity. Plasma CK is used frequently in human medicine to study skeletal muscle diseases, and plasma CK activities may be markedly affected by skeletal muscle damage either by simple intramuscular injections or in toxic myopathies. However, plasma values often do not appear to change significantly in laboratory animals, although there may be clinical and histological evidence of muscle injury: again, this may be a matter of timing of the blood sample collections for the appropriate half-life of the species being examined. Exercise may increase the efflux of muscle enzymes – CK, LDH and AST – into the plasma, but these exercise effects are variable and can be altered by the duration and form of exercise in animals (Loegering, 1974; Sanders and Bloor, 1975).

There are several examples where plasma enzyme changes associated with myotoxicity have been described. In rats, chlorpromazine given intramuscularly but not intraperitoneally, increased serum CK levels (Meltzer *et al.*, 1970). Plasma LDH and CK increases in rabbits, dosed orally with the lipid lowering agent simvastatin, were associated with severe lesions of skeletal muscle; this could be due to the hydrophobic simvastatin permeating the plasma membrane (Fukami *et al.*, 1993). Owen and colleagues (1994) reported enzyme changes occurring with myopathy in marmosets following the administration of an inhibitor of 3-hydroxy-3-methyl-glutaryl-coenzyme A (HMGCoA) – another hypolipidaemic compound.

Serum CK changes, associated with histological evidence of muscle cell necrosis, have been demonstrated in dogs following single intramuscular injections of benzoctamin, diazepam and pethidine (Gloor *et al.*, 1977). Steiness *et al.* (1978) reported changes for serum CK in rabbits and pigs following single intramuscular injections of lidocaine, diazepam and digoxin; the same authors also reported serum CK activity increased with successive blood sample collections from untreated rabbits.

With intramuscular injections of tetracycline in rats, Swain and colleagues (1994) found serum CK levels peaked at 2 h post-injection and returned to below baseline values at 24 h. The serum CK levels increased more than threefold in rhesus monkeys 15 min after injection with ketamine (Bennett *et al.*, 1992). The results of these two studies emphasize the rapid response of CK following intramuscular injection.

Sometimes, the increase of plasma enzyme activities and muscle necrosis may not be due to the drug itself. Several other factors include osmolality, hydrogen ion concentration (pH) and volume of the solution administered; the reaction may also be due to the excipient used in the drug formulation, or it may differ with slow release formulations. Surber and Dubach (1989) demonstrated varied responses in rats given differently formulated multivitamin preparations. Elevated serum CK levels have also been reported in experimentally induced myositis (Esiri and Maclennan, 1974), and following experimental intestinal infarction in rats (Roth *et al.*, 1989) and dogs (Graeber *et al.*, 1981).

References

Balazs, T. & Bloom, S. (1982) Cardiotoxicity of adrenergic, bronchodilator and vasodilating antihypertensive drugs. In Van Stee, E. W. (ed.), *Cardiovascular Toxicity*, pp. 199–200. New York: Raven Press.

Barrett, R. J., Harleman, H. & Joseph, E. C. (1988) The evaluation of HBDH and LDH isoenzymes in cardiac cell necrosis of the rat. *Journal of Applied Toxicology*, **8**, 233–8.

Bennett, J. S., Gossett, K. A., McCarthy, M. P. & Simpson, E. D. (1992) Effects of ketamine hydrochloride on serum biochemical and hematologic variables in rhesus monkeys (*Macaca mulatta*). *Veterinary Clinical Pathology*, **21**, 15–18.

Bhargava, A. S., Preus, M., Khater, A. R. & Gunzel, P. (1990) Effect of Iloprost on serum creatine kinase and lactate dehydrogenase isoenzymes after isoprenaline-induced cardiac damage in rats. *Arzneimittel-Forschung*, **40**, 248–52.

Braun, S. (1992) Isoformen der Creatinkinase-Isoenzyme. *Klinische Labor*, **38**, 549–54.

Czarnecki, C. M. (1984) Animal models of drug-induced cardiomyopathy. *Comparative Biochemistry and Physiology*, **79C**, 9–14.

Dimitrijevic, N., Vulovic, D., Vlajnic, R., Protic, S., Dordevic-Denic, G. & Jovanovic, B. (1993) Total activity and isoenzymic patterns of lactate dehydrogenase (LDH) in various rat tissues. *Acta Veterinaria (Beograd)*, **43**, 87–94.

Esiri, M. M. & Maclennan, I. C. M. (1974) Experimental myositis in rats. *Clinical and Experimental Immunology*, **17**, 139–50.

Evans, G. O. (1991) Biochemical assessment of cardiac function and damage in animal species. *Journal of Applied Toxicology*, **11**, 15–21.

Fayolle, P., Lefebvre, H. & Braun, J. P. (1992) Effects of incorrect venepuncture on plasma creatine-kinase activity in dog and horse. *British Veterinary Journal*, **148**, 161–2.

Fukami, M., Maeda, N., Fukushige, J., Kogure, Y., Shimada, Y., Ogawa, T. & Tsujita, Y. (1993) Effects of HMG-CoA reductase inhibitors on skeletal muscles of rabbits. *Research in Experimental Medicine*, **193**, 263–73.

Garbus, G. B., Highman, B. & Altland, P. D. (1967) Alterations in serum enzymes and isoenzymes in various species induced by epinephrine. *Comparative Biochemistry and Physiology*, **22**, 507–16.

Gloor, H. O., Vorburger, C. & Schadelin, J. (1977) Intramuskulare Injektionen und Serumkreatinphosphokinase-aktivitat. *Schweizerische Medizinische Wochenschrift*, **107**, 948–52.

Graeber, G. M., Cafferty, P. J., Rearden, M. J., Curley, C. P., Ackerman, N. B. & Harmon, J. W. (1981) Changes in serum creatine kinase CPK and its isoenzymes caused by experimental ligation of the superior mesenteric artery. *Annals of Surgery*, **193**, 499–505.

HAYASHI, T. & NOTKINS, A. L. (1994) Clearance of LDH-5 from the circulation of inbred mice correlates with binding of macrophages. *International Journal of Experimental Pathology*, **75**, 165–8.

KARLSSON, K. B. & LARSSON, G. B. (1971) Lactic and malic dehydrogenase and their multiple molecular forms in the mongolian gerbil as compared with the rat, mouse and rabbit. *Comparative Biochemistry and Physiology*, **40B**, 93–108.

KATUS, H. A., SCHEFFOLD, T., REMPPIS, A. & ZEHLEIN, J. (1992) Proteins of the troponin complex. *Laboratory Medicine*, **23**, 311–17.

KIM, S. N., WATKINS, J. R., JAYASEKARA, J., ANDERSON, J. A., FITZGERALD, J. E. & DE LA IGLESIA, F. A. (1985) Cardiotoxicity study of amsacrine in rats. *Toxicology and Applied Pharmacology*, **77**, 369–73.

LANDT, Y., VAIDYA, H. C., PORTER, S. E., DIETZLER, D. N. & LADENSON, J. H. (1989) Immunoaffinity purification of creatine kinase-MB from human, dog and rabbit heart with use of a monoclonal antibody specific for CK-MB. *Clinical Chemistry*, **35**, 985–9.

LANG, H. (1981) *Creatine kinase isoenzymes*. Berlin: Springer-Verlag.

LARAGH, J. H. (1985) Atrial natriuretic hormone. The renin–aldosterone axis and blood pressure–electrolyte homeostasis. *New England Journal of Medicine*, **313**, 1330–40.

LINDENA, J., DIEDERICHS, F., WITTENBERG, H. & TRAUTSCHOLD, I. (1986) Kinetic of adjustment of enzyme catalytic concentrations in the extracellular space of the man, the dog, and the rat. *Journal of Clinical Chemistry and Clinical Biochemistry*, **24**, 61–71.

LOEGERING, D. J. (1974) Effect of swimming and treadmill exercise on plasma enzyme levels in rats. *Proceedings of the Society for Experimental Biology and Medicine*, **147**, 177–80.

MAEJIMA, K. & NAGASE, S. (1991) Effects of starvation and refeeding on the circadian rhythms of hematological and clinico-biochemical values and water intake of rats. *Experimental Animals*, **40**, 389–93.

MALOUF, N. M., MCMAHON, D., OAKELEY, A. E. & ANDERSON, P. A. W. (1992) A cardiac troponin T epitope conserved across phyla. *Journal of Biological Chemistry*, **267**, 9269–74.

MARKERT, M. L. & WHITT, G. (1975) Evolution of a gene. *Science*, **189**, 102–14.

MATSUZAWA, T. & ISHIKAWA, A. (1993) The effect of preanalytical conditions on lactate dehydrogenase and creatine kinase activities in the rat. *Comparative Haematology International*, **3**, 214–19.

MATSUZAWA, T., NOMURA, M. & UNNO, T. (1993) Clinical pathology reference ranges of laboratory animals. *Journal of Veterinary Medicine and Science*, **55**, 351–62.

MCBRIDE, J. H., RODGERSON, D. O. & HILBORNE, L. H. (1990) Human, rabbit, bovine and porcine creatine kinases isoenzymes are glycoproteins. *Journal of Clinical Laboratory Analysis*, **4**, 196–8.

MELTZER, Y. & GUSCHWAN, A. (1971) Effect of isoproterenol on rat plasma creatine phosphokinase activity. *Archives of International Pharmacodynamics*, **194**, 141–6.

MELTZER, H. Y., MROZAK, S. & BOYER, M. (1970) Effect of intramuscular injections on serum creatine phosphokinase activity. *American Journal of Medical Sciences*, **259**, 42–8.

MYERS, C. E. (1988) Role of iron in anthracycline action. In Hacker, M. P., Lazo, J. S. & Tritton, T. R. (eds), *Organ Directed Toxicities of Anticancer Drugs*, pp. 17–30. Boston: Martinus Nijhoff.

OLSON, H. M. & CAPEN, C. C. (1977) Subacute cardiotoxicity of adriamycin in the rat. Biochemical and ultrastructural investigations. *Laboratory Investigation*, **37**, 386–94.

OWEN, K., PICK, C. R., LIBRETTO, S. E. & ADAMS, M. J. (1994) Toxicity of a novel HMG-CoA reductase inhibitor in the common marmoset (*Callithrix jacchus*). *Human and Experimental Toxicology*, **13**, 357–68.

POLDASEK, S. J. & MCPHERSON, R. A. (1989) Streptokinase binds to lactate dehydrogenase subunit-M, which shares an epitope with plasminogen. *Clinical Chemistry*, **35**, 69–73.

ROTH, M., JAQUET, P.-Y. & ROHNER, A. (1989) Increase of creatine kinase and lactate dehydrogenase in the serum of rats submitted to experimental intestinal infarction. *Clinica Chimica Acta*, **183**, 65–70.

SADEK, S. E. & PFITZER, E. A. (1975) Correlation of histologic and serum changes of isoproterenol-induced myocardial lesions in dogs. *Toxicology and Applied Pharmacology*, **33**, 156–7.

SAGGIN, L., GORZA, L., AUSONI, S. & SCHIAFFINO, S. (1990) Cardiac troponin T in developing regenerating and denervated rat skeletal muscle. *Development*, **110**, 547–54.

SANDERS, T. M. & BLOOR, C. M. (1975) Effect of endurance exercise on serum enzyme activities in the dog, pig and man. *Proceedings of the Society for Experimental Biology and Medicine*, **148**, 823–8.

SCHULTZE, A. E., GUNAGA, K. P., WAGNER, J. G., HOORN, C. M., MOOREHEAD, W. R. & ROTH, R. A. (1994) Lactate dehydrogenase activity and isoenzyme patterns in tissues and broncholalveolar lavage fluid from rats treated with monocrotaline pyrrole. *Toxicology and Applied Pharmacology*, **126**, 301–10.

STEINESS, E., RASMUSSEN, F., SVENDSEN, O. & NIELSEN, P. (1978) A comparative study of serum creatine phosphokinase (CPK) activity in rabbits, pigs and humans after intramuscular injection of local damaging drugs. *Acta Pharmacology et Toxicology*, **42**, 357–64.

SURBER, C. & DUBACH, U. C. (1989) Tests for local toxicity of intramuscular drug preparations. Comparison of *in vivo* and *in vitro* findings. *Arzneimittel-Forschung*, **39**, 1586–9.

SWAIN, R., WILLIAMS, G. & COCHRAN, J. (1994) Elevation and clearance of serum creatine kinase in Fischer rats. *Clinical Chemistry*, **40**, 990.

VOSS, E., MURAKAMI, M. A., GERNERT, A., JOHNSTON, R., SHARKEY, S. & APPLE, F. (1994) Utility of serum cardiac troponin T and CK-MB for assessing canine myocardial infarct size. *Clinical Chemistry*, **40**, 1040–1.

YASMINEH, W. G., PYLE, R. B., HANSON, N. Q. & HULTMAN, B. K. (1976) Creatine kinase isoenzymes in baboon tissues and organs. *Clinical Chemistry*, **22**, 63–6.

ZBINDEN, G., DECAMPEENERE, D. & BAURAIN, R. (1991) Preclinical assessments of the cardiotoxic potential of anthracycline antibiotics: N-L-leucyl-doxorubicin. *Archives of Toxicology*, Suppl. **14**, 107–17.

11

Assessment of Neurotoxicity

M. D. STONARD

11.1 Introduction

The structural and functional diversity of the nervous system does not lend itself readily to any single analyte or macromolecule which may be measured as an indicator of early damage to, or the functional integrity of, this complex organ system. Sampling of body fluids such as blood and urine has not proved generally useful in the diagnosis of damage and dysfunction of the nervous system. Collection of body fluids such as cerebrospinal fluid is a technically difficult invasive procedure and is not performed routinely. Thus the clinical biochemist must resort to the use of nervous system tissue in which to measure the relevant analyte or macromolecule. Relatively few marker molecules of diagnostic significance have been described, but with the increasing use of molecular biology techniques it is anticipated that the number will grow substantially.

The complexity of the nervous system is known to reflect functional heterogeneity in which metabolic pathways can be localized to discrete areas of the brain. Thus the removal of nervous system tissue from experimental animals at necropsy and subsequent processing to prepare a homogenate will provide a measure of the total activity or concentration of an analyte. It will not indicate whether the activity or concentration of the analyte varies from region to region. Furthermore, if it can be shown that experimental treatment with a foreign compound reduces the activity or concentration of an analyte in a homogenate, this measurement must be followed by further measurements in specific regions of the brain to determine whether damage is selective. This can be illustrated by reference to the enzyme cholinesterase, which is involved in neurotransmission.

11.2 Cholinesterases

Two principal types of cholinesterase enzyme have been identified:

1 Acetylcholinesterase (AChE; EC 3.1.1.7). This enzyme is found both in the brain and erythrocytes and is characterized by its preferential affinity for

acetylcholine as substrate. It is also referred to as 'true' cholinesterase. The enzyme is known to exist in different polymorphic forms (Skau, 1985).

2 Cholinesterase found in plasma/serum and several other tissues, e.g. liver (see Chapter 2), exhibits both species specificity (Myers, 1953) and tissue heterogeneity (Ecobichon and Comeau, 1972; Unakami *et al.*, 1987) and shows less substrate specificity than AChE. It is also referred to as 'pseudo-' or 'non-specific' cholinesterase (ChE; EC 3.1.1.8).

The mechanism of action of AChE involves an initial step in which the enzyme forms a complex (the Michaelis complex) with the substrate, acetylcholine, which then acetylates the enzyme with the release of choline. This complex is reversible such that hydrolysis leads to the release of acetic acid and the regeneration of the de-acetylated enzyme. This enzyme is vulnerable to attack by certain classes of chemical compound. Organophosphorus esters, which are used in agriculture as insecticides, react with the enzyme in a manner analogous to that of acetylcholine, and phosphorylate the enzyme. For many of these chemicals, the phosphorylated enzyme complex is stable and only slowly reversible. Similarly, esters of carbamic acid are used as insecticides and can interact with AChE to form a carbamylated enzyme complex which is more readily reversible than a phosphorylated complex. These classes of compound can also interact with ChE, and the susceptibility of the two classes of enzyme varies from compound to compound.

Not all total cholinesterase activity in the nervous system can be accounted for by AChE. It is estimated that some 15% of total activity is due to the non-specific ChE enzyme in certain areas of the white matter (Ecobichon and Joy, 1982). The physiological role of the non-specific ChE is a matter of conjecture. It is synthesized in the liver but its natural substrate requirements are unclear. Thus any inhibition of this enzyme by organophosphorus or carbamate esters can only be interpreted as evidence of absorption of these chemicals into the body without any assessment of the health significance of the finding. In contrast to the non-specific enzyme, AChE has an unequivocal role as one of the two key enzymes involved in acetylcholine neurotransmission. Whilst the enzyme choline acetyltransferase (EC 2.3.1.6) is key to the synthesis of acetylcholine, AChE is essential to the inactivation of acetylcholine and is localized accordingly, e.g. at neuromuscular end-plates. Any interference with this enzyme at critical sites in the nervous system will cause, therefore, cholinergic nerve impulses to be perpetuated leading to possible nerve or muscle damage. Organophosphorus anticholinesterase agents are generally more toxic to the young animal in comparison with the adult (Lu *et al.*, 1965; Pope *et al.*, 1991).

For obvious reasons, brain AChE measurements can only be undertaken in animal species at necropsy. Therefore, in humans involved in the manufacture or spraying of organophosphorus or carbamate ester insecticides, measurements of erythrocyte and/or plasma cholinesterase activities are employed to provide an index of compound absorption and potential to have an adverse effect on the nervous system (Wills, 1972). The relative potency of the chemical towards the erythrocyte and plasma enzymes can be evaluated, and the more sensitive enzyme used to monitor exposure. In practice, baseline measurements of both enzymes are established for the individual prior to chemical exposure and one or both enzymes measured at subsequent intervals during exposure. The frequency of monitoring will depend on the level of anticipated exposure and the characteristics of the interaction with AChE, i.e. the slower the rate of recovery of the inhibited enzyme, the greater the frequency of monitoring. In occupational practice, therefore, a comparison of the enzyme activities during exposure with those obtained pre-exposure can be used as one measure to monitor the hazard to health.

For safety evaluation studies involving new organophosphorus and carbamate esters which are being developed as candidate insecticides, it is essential to measure both cholinesterase activities in blood and brain. Regulatory guidelines usually dictate the frequency of blood measurements in specific study types, e.g. chronic toxicity with brain measurements made in all or some of the animals at necropsy, depending on the species (US EPA, 1984; OECD, 1987). The dog and rat are the usual species of choice. While concurrent control groups are a feature of both dog and rat studies, and are used as the point of comparison with treated groups, pretreatment baseline values may be established for dogs, where the group sizes are smaller than for rats. This is essential, since in humans, and probably in dogs, the variation in both cholinesterase enzymes between individuals exceeds the variation between successive determinations in the same individual (Callaway *et al.*, 1951; Gage, 1967). The percentage difference from the mean baseline value which achieves statistical significance is, of course, dependent upon which enzyme is being measured and upon the number of baseline measurements (Hayes, 1982). Various sources of variability (including analytical) confound the interpretation of cholinesterase data. For animal studies it has been usual to assume that up to 20% inhibition difference from the control is within normal variability for both brain and blood cholinesterase measurements. This is based upon a combination of analytical, inter- and intravariability seen in the human population (Callaway *et al.*, 1951; Gage, 1967; Hackathorn *et al.*, 1983).

All measurements of cholinesterase are based upon the hydrolysis of an ester by the enzyme. A detailed review of the biochemical measurement and significance of cholinesterase is given by Wills (1972). Histochemical methods (Koelle, 1954; Storm-Mathison and Blackstad, 1964) have been used to demonstrate the distribution of AChE in the nervous system of post-mortem rat and human brain (Biegon and Wolff, 1985), and the rhesus monkey (Mesulam *et al.*, 1986). Several biochemical methods have been developed for the measurement of enzyme activity (see review by Whittaker, 1986). The enzyme reactions for both enzymes are sensitive to pH, temperature and substrate concentration and all of these factors must be considered in order to optimize the assay for the relevant species. A continuous kinetic rate measurement of the formation of product is the usually preferred method of choice in most laboratories. There are basically four different procedures: 1. potentiometric (pH change), 2. radiometric, 3. manometric and 4. colorimetric. The potentiometric technique involves the measurement of pH, which falls as acetic acid released by hydrolysis of the acetylated enzyme occurs. The method lacks sensitivity and is not widely used (Michel, 1949). The radiometric technique relies on the release of radiolabelled acetic acid from radiolabelled acetylcholine. The method is highly sensitive but requires a scintillation counter (Winteringham and Disney, 1964; Johnson and Russel, 1975). The manometric method is based upon the release of carbon dioxide evolved from the reaction of the bicarbonate in the system with the acid formed by hydrolysis of the ester (Augustinsson, 1948).

The most widely used technique is the colorimetric assay developed by Ellman *et al.*, 1961. This is based upon the use of acetylated thiocholine(s) which are hydrolysed to thiocholine, and which combined with 5,5-dithio-bis-2-nitrobenzoic acid (DTNB) produces a yellow colour. The yellow colour has a large extinction coefficient and is therefore a highly sensitive method. There are significant pitfalls associated with the measurement of cholinesterase. The time between blood sampling and assay should be minimized to prevent further inhibition of the enzyme by free inhibitor. Equally, if spontaneous reactivation of the inhibited enzyme occurs relatively rapidly, the incubation time for the assay must be minimized following dilution of the blood sample. If these conditions are not controlled, then the degree of inhibition of cholinesterase may be overestimated or underestimated. Thus for the generation of meaningful data, it is essential that the time which elapses between tissue sampling and assay (including the assay incubation period) should be kept to a minimum and samples cooled and the pH reduced if necessary (Wilhelm and Reiner, 1973).

11.3 Neurotoxic Esterase (NTE)

The identification of a specific esterase activity amongst all of the esterase activity in the nervous system has provided the basis for a selective screening method for acute delayed neurotoxicity. The development of an assay for this discrete enzyme activity owes much to the pioneering work of Johnson (1974, 1977). As a chemical class, most organophosphorus esters are direct inhibitors or are converted to inhibitors of AChE. A relatively small proportion of these compounds has the ability to produce a delayed neurotoxicity referred to as organophosphate-induced delayed polyneuropathy (OPIDP). Following a single dose of one of these compounds, clinical signs of neurotoxicity do not develop until 6–14 days after dosing. The hen which is a particularly sensitive species develops a flaccid paralysis. Histopathological examination of the nervous system has revealed damage to the peripheral nerves and spinal cord but not the higher brain. It has been shown that the initial step in the process which leads to delayed neurotoxicity is the interaction of the compound with an enzyme, neurotoxic (or neuropathy target) esterase (NTE). This interaction occurs within hours of dosing and clearly precedes the development of the adverse clinical signs.

Several cases of OPIDP have been documented in humans, who are considered, like the hen, to be a sensitive species. However, not all species are susceptible to OPIDP; cats and farm species (cows, sheep) are susceptible whereas rats, mice, rabbits, guinea pigs and hamsters have not exhibited a consistent delayed neurotoxic response to the classical neuropathic OP, namely tri-ortho cresyl phosphate. Several avian species have also been shown to be sensitive to OPIDP. The adult hen has been widely accepted as a suitable and sensitive species for evaluating the potential hazards posed to man by organophosphorus esters. The use of the hen as the species of choice is reflected in the regulatory guidelines issued by the US EPA (1991) and the OECD (1987), the latter having undergone recent revision.

The basis for an assay of neurotoxic esterase activity evolved from studies to delineate the mechanism of action of OPIDP. Hen brain contains substantial esterase activity, and with the use of both neurotoxic and non-neurotoxic OPs, it was possible to identify a small proportion of esterase activity that was susceptible to inhibition only by OP esters which caused delayed neurotoxicity. By the use of phenyl valerate or phenyl phenyl acetate as substrate, it was shown that approximately 6% of the total esterase activity was associated with the development of delayed neurotoxicity. The technical aspects of the NTE assay, including the preparation of the reagents, have been addressed (Johnson, 1977, 1982; Johnson and Richardson, 1983). This differential assay uses phenyl valerate as the substrate and two inhibitors, mipafox and paraoxon, as the neurotoxic and non-neurotoxic agents respectively.

Inhibition of NTE is not confined to OPs but also is seen with carbamate esters. However, in contrast to OPs, inhibition by carbamate esters does not lead to delayed neurotoxicity. The explanation for the difference in response between OPs and carbamate esters lies in the ability of some OPs to undergo a second reaction. The phosphorylation of the esterase active site by the OP can, in some instances, leave the phosphoryl moiety attached to the active site with a negative charge. A similar reaction cannot occur with carbamate esters. This second stage is referred to as 'aging' and is an obligatory step in the development of delayed neurotoxicity. There are, therefore, at least two stages in the development of acute delayed neurotoxicity. First, the enzyme must be inhibited by at least 70–80% to have the potential to cause delayed neurotoxicity, and second the phosphorylated enzyme must be capable of undergoing 'aging'. The first stage of the process can be measured reliably using the established procedure for NTE. However, the second stage is technically more difficult to measure, and is currently not available as a routine procedure. Nevertheless, there is a wealth of information available which can be used to predict whether a compound will undergo 'aging' or not.

11.4 Glial Fibrillary Acidic Protein (GFAP)

In the central nervous system, astrocytes undergo both hypertrophy and proliferation in response to neuronal cell injury, a process referred to as reactive gliosis. This process is a prominent reaction of astrocytes adjacent to and extending far beyond the site of injury. Astrocytes are activated by a variety of insults, including physical, chemical, viral and bacterial. These cells contain a unique cytoskeletal protein in their intermediate filaments known as glial fibrillary acidic protein (GFAP). Astrocytes, which are the most prolific glial cells, provide structural support for neurones and play a key role in the repair of the central nervous system after damage. GFAP has been isolated and characterized from astrocytes (Eng *et al.*, 1971). Reactive gliosis is characterized by extensive synthesis of GFAP (Eng, 1985). Evidence from studies with rat optic nerve astrocyte cultures suggests intermediate filaments are important for structural stability of astrocytes (Trimmer *et al.*, 1982). The relatively slow metabolic turnover rate for GFAP is consistent with such a structural role (Smith *et al.*, 1984; De Armand *et al.*, 1986).

The potential to detect neuronal cell injury by measurement of GFAP in brain tissue has been investigated in several laboratories. Immunohistochemical procedures are available to provide a semi-quantitative estimation by light microscopy (Eng and De Armand, 1983). Validation of GFAP as a biochemical indicator of neuronal cell injury required the development of a sensitive and reliable method for brain tissue. Initially, a detergent-based solid-phase radioimmunoassay was developed (O'Callaghan and Miller, 1985). Significant improvements to the original assay have been made with both a slot-immunobinding assay (Brock and O'Callaghan, 1987), and more recently a novel sandwich ELISA (O'Callaghan, 1991). Essentially similar results were obtained with the polyclonal slot blot, monoclonal slot blot and the sandwich ELISA assays. The latter assay offers significant advantages over the slot-binding assays by having fewer steps, using less reagents, and is colorimetric not radiometric.

In response to the requirement by the US EPA to measure GFAP in neurotoxicity studies (US EPA, 1991), several laboratories have introduced a quantitative method for measurement in both whole brain and discrete regions of the brain (Dickens *et al.*, 1993). Age-related increases in GFAP per unit of protein were demonstrated in normal rats between the ages of 6 and 39 weeks, with large increases observed consistently in the striatum/thalamus. This increase in GFAP in older animals may reflect an accumulation of intermediate filaments within the astrocytes of the aging brain, or an age-related astrocytic hypertrophy/hyperplasia, all of which may represent a response to neuronal loss which occurs with age. Trimethyltin has been used to produce selective injury to the central nervous system. When administered as a single oral or i.v. dose to rats, trimethyltin caused increases in GFAP in selective regions of the brain, particularly the striatum and cerebral cortex (Dickens *et al.*, 1993), and hippocampus (O'Callaghan, 1991).

In summary, there are relatively few analytes which can be used as indices of damage and dysfunction to the nervous system. At present, with the exception of blood cholinesterases, they all require the measurements to be made in the target tissue or discrete regions of the nervous system.

References

AUGUSTINSSON, K. B. (1948) Cholinesterases, a study in comparative enzymology. *Acta Physiologica Scandinavia*, 15, Suppl. **52**, 1–192.

BIEGON, A. & WOLFF, M. (1985) Quantitative histochemistry of acetylcholinesterase in rat and human brain postmortem. *Journal of Neuroscience Methods*, **16**, 39–45.

BROCK, T. O. & O'CALLAGHAN, J. P. (1987) Quantitative changes in the synaptic vesicle proteins, synapsin 1 and p38, and the astrocyte specific protein, GFAP, are associated with chemical-induced injury to the rat central nervous system. *Journal of Neuroscience*, **7**, 931–42.

CALLAWAY, S., DAVIES, D. R. & RUTLAND, J. P. (1951) Blood cholinesterase levels and a range of personal variation in a healthy adult population. *British Medical Journal*, **2**, 812.

DE ARMAND, S. J., LEE, Y.-L., KRETZSCHMAR, H. A. & ENG, L. F. (1986) Turnover of glial filaments in mouse spinal cord. *Journal of Neurochemistry*, **47**, 1749–53.

DICKENS, A. D., LEA, L. J., ROBINSON, J. A. & SLACK, I. (1993) Measurement of glial fibrillary acidic protein (GFAP) in the rat and validation of its use as a biochemical marker of neurotoxicity. *Comparative Haematology International*, **3**, 57.

ECOBICHON, D. J. & COMEAU, M. (1972) Pseudocholinesterases of mammalian plasma: physicochemical properties and organophosphate inhibition in eleven species. *Toxicology and Applied Pharmacology*, **24**, 92–100.

ECOBICHON, D. J. & JOY, R. M. (1982) *Pesticides and Neurological Disease*. Boca Raton, Florida: CRC Press.

ELLMAN, G. L., COURTNEY, K. D., ANDRES, V. & FEATHERSTONE, R. M. (1961) A new and rapid colorimetric determination of acetylcholinesterase activity. *Biochemical Pharmacology*, **7**, 88–95.

ENG, L. F. (1985) Glial fibrillary acidic protein: the major protein of glial intermediate filaments in differentiated astrocytes. *Journal of Neuroimmunology*, **8**, 203–14.

ENG, L. F. & DE ARMAND, S. J. (1983) Immunochemistry of the glial fibrillary acidic protein. In Zimmerman, H. M. (ed.), *Progress in Neurobiology*, Vol. 5, pp. 19–39. New York: Raven Press.

ENG, L. F., VANDERHAEGEN, J. J., BIGNAMI, A. & GERSTL, B. (1971) An acidic protein isolated from fibrous astrocytes. *Brain Research*, **28**, 351–4.

GAGE, J. C. (1967). The significance of blood cholinesterase activity measurements. *Residue Reviews*, **18**, 159–73.

HACKATHORN, D. R., BRINKMAN, W. J., HATHAWAY, T. R., TALBOT, T. D. & THOMPSON, L. R. (1983) Validation of whole blood method for cholinesterase monitoring. *American Industrial Hygiene Association Journal*, **44**, 547–51.

HAYES, W. J. (1982) *Pesticides Studied in Man*. Baltimore: Williams & Wilkins.

JOHNSON, C. D. & RUSSEL, P. L. (1975) A rapid, simple radiometric assay for cholinesterase, suitable for multiple determinations. *Analytical Biochemistry*, **64**, 229–38.

JOHNSON, M. K. (1974) The primary biochemical lesion leading to the delayed neurotoxic effect of some organophosphorus esters. *Journal of Neurochemistry*, **23**, 785–98.

(1977) Improved assay of neurotoxic esterase for screening organophosphates for delayed neurotoxicity potential. *Archives of Toxicology*, **37**, 113–15.

(1982) The target for initiation of delayed neurotoxicity by organophosphorus esters: biochemical studies and toxicological applications. In Hodgson, E., Bend, J. R. & Philpot, R. M. (eds), *Reviews in Biochemistry and Toxicology 4*, pp. 141–212. New York: Elsevier.

JOHNSON, M. K. & RICHARDSON, R. J. (1983) Biochemical endpoints: neurotoxic esterase assay. *Neurotoxicology*, **4**, 311–20.

KOELLE, G. B. (1954) The histochemical localization of cholinesterases in the central nervous system of the rat. *Journal of Comparative Neurology*, **100**, 211–21.

LU, F. C., JESSUP, D. C. & LAVALLEE, A. (1965) Toxicity of pesticides in young versus adult rats. *Food and Cosmetic Toxicology*, **3**, 591–6.

MESULAM, M-M., VOLICER, L., MARQUIS, J. K., MUFSON, E. J. & GREEN, R. C. (1986) Systematic regional differences in the cholinergic innervation of the primate cerebral cortex: distribution of enzyme activities and some behavioral implications. *Annals of Neurology*, **19**, 144–51.

MICHEL, H. O. (1949) An electrometric method for the determination of red blood cell and plasma cholinesterase activity. *Journal of Laboratory and Clinical Medicine*, **34**, 1564–8.

MYERS, D. K. (1953) Studies on cholinesterase 9. Species variation in the specificity pattern of the pseudo-cholinesterases. *Biochemical Journal*, **55**, 67–75.

O'CALLAGHAN, J. P. (1991) Quantification of glial fibrillary acidic protein: comparison of slot-immunobinding assays with a novel sandwich ELISA. *Neurotoxicology and Teratology*, **13**, 275–81.

O'CALLAGHAN, J. P. & MILLER, D. B. (1985) Cerebellar hypoplasia in the Gunn rat is associated with quantitative changes in neurotypic and gliotypic proteins. *Journal of Pharmacology and Experimental Therapeutics*, **234**, 522–33.

OECD (1987) Guidelines for testing of chemicals. Section 4 Health Effects.

POPE, C. N., CHAKRABORTI, T. K., CHAPMAN, M. L., FARRAR, J. D. & ARTHUN, D. (1991) Comparison of *in vivo* cholinesterase inhibition in neonatal and adult rats by three organophosphorothioate insecticides. *Toxicology*, **68**, 51–61.

SKAU, K. A. (1985) Acetylcholinesterase molecular forms in serum and erythrocytes of laboratory animals. *Comparative Biochemistry and Physiology*, **80C**, 207–10.

SMITH, M. E., PERRET, V. & ENG, L. F. (1984) Metabolic studies *in vitro* of the CNS cytoskeletal proteins: synthesis and degradation. *Neurochemical Research*, **9**, 1493–1507.

STORM-MATHISON, J. & BLACKSTAD, T. W. (1964) Cholinesterase in the hippocampal region. Distribution and relation to architectonics and afferent systems. *Acta Anatomy*, **56**, 216–53.

TRIMMER, P. A., REIER, P. J., OH, T. H. & ENG, L. F. (1982) An ultrastructural and immunocytochemical study of astrocytic differentiation *in vitro*. Changes in the composition and distribution of the cellular cytoskeleton. *Journal of Neuroimmunology*, **2**, 235–60.

UNAKAMI, S., SUZUKI, S., NAKANISHI, E. *et al.* (1987) Comparative studies on multiple forms of serum cholinesterase in various species. *Experimental Animals*, **36**, 199–204.

US EPA (1984) Pesticide assessment guidelines, Subdivision F, Hazard evaluation: Human and domestic animals.

(1991) Pesticide assessment guidelines, Subdivision F, Hazard evaluation: Human and domestic animals, addendum 10, Neurotoxicity, Series 81, 82 and 83.

WHITTAKER, M. (1986) *Cholinesterase. Monographs in Human Genetics*, Vol. II. Basel: Karger.

WILHELM, K. & REINER, E. (1973) Effect of sample storage on human blood cholinesterase activity after inhibition by carbamates. *Bulletin of the World Health Organization*, **48**, 235–8.

WILLS, J. H. (1972) The measurement and significance of changes in the cholinesterase activities of erythrocytes and plasma in man and animals. *CRC Critical Reviews in Toxicology*, **1**, 153–202.

WINTERINGHAM, F. P. W. & DISNEY, R. W. (1964) A simple method of estimating blood cholinesterase. *Laboratory Practice*, **13**, 739–40.

12

Electrolyte and Fluid Balance

M. J. YORK & G. O. EVANS

Depending on their electrical charge, electrolytes exist as anions or cations, and electrical neutrality requires an equal number of anions and cations. In the body, the electrolytes fulfil a number of vital roles in maintaining fluid balance and pH (hydrogen ion concentration), membrane potentials, muscular functions, nervous conduction and they serve as cofactors in many enzyme-mediated reactions.

Many laboratories confine electrolyte measurements in general toxicological studies to plasma sodium (Na^+), potassium (K^+), calcium (Ca^{++}), and urinary sodium and potassium. Magnesium (Mg^{++}) and anions such as inorganic phosphate, chloride (Cl^-), bicarbonate (HCO_3^-) and lactate are measured less frequently, although the use of combination triple-ion (Na^+, K^+ and Cl^-) selective electrodes sometimes results in an unwanted measurement of chloride. Urinary sodium and potassium are often measured to meet regulatory requirements, where renal toxicity or marked alterations of fluid balance are suspected. Some molecules such as amino acids which are also ionized are rarely measured in regulatory toxicology studies. Proteins also carry an electrical charge but these are considered separately in Chapter 11.

For most animals, the total body water is approximately 60% of body weight, and it can be separated into two interlinked fluidic compartments: intracellular (ICF) and extracellular (ECF). These two compartments are normally in an osmotic equilibrium governed by the osmotically active molecules in each compartment and with water freely diffusible between the ICF and ECF. However, the electrolyte constituents of the ICF and ECF are markedly different.

The ECF consists of several body fluids including plasma, interstitial fluid, lymph and transcellular fluids including the secretions of the gastrointestinal tract. On rare occasions, it is necessary to analyse the constituents of some of these fluids where excessive accumulation occurs.

Although both intake and excretion of fluids and electrolytes may vary considerably in toxicological studies, ionic concentrations in the body fluids often appear to be well controlled with minimum perturbations of the common plasma electrolyte measurements but these concentrations may change following severe toxicity. It is not uncommon to find small but statistically significant changes of electrolyte measurements for either individuals or groups of animals which do not appear to be dose related nor correlate with histopathological findings. In some species, e.g. monkeys, the plasma electrolyte concentrations may be highly variable.

12.1 Endocrine Control of Electrolyte and Fluid Balance

Regulatory mechanisms involve several hormones which include aldosterone, arginine vasopressin, atrial natriuretic factor and the renin–angiotensin axis. These hormones have differing sites of synthesis and actions (Laragh, 1985).

12.1.1 *Arginine Vasopressin (AVP)*

Also called antidiuretic hormone (ADH), this nonapeptide is synthesized in the hypothalamus and secreted via the posterior pituitary gland. The hormone is released in response to changes of osmolality and plasma sodium, which is a major determinant of plasma osmolality (Robertson, 1987; see later). In the renal collecting tubules, AVP acts on adenylate cyclase generating adenosine monophosphate and protein kinases, which in turn alter the permeability of the renal collecting ducts and distal tubules to water. The threshold for AVP release varies with plasma osmolality values in the different species. AVP release is also stimulated by a depletion of the body's effective circulating fluid volume, and AVP acts as an arterial vasoconstrictor thus altering blood pressure (Mohr and Richter, 1994). AVP is not detectable in the plasma of Brattleboro rats (Horn *et al.*, 1985).

12.1.2 *Renin–Angiotensin Axis*

The proteolytic enzyme renin is synthesized in the juxta-glomerular apparatus of the glomerular afferent arterioles. Renin is secreted in response to reduced renal perfusion and excessive loss of sodium (Oparil and Haber, 1974a, 1974b), and it cleaves angiotensinogen, an $\alpha 2$ globulin produced by the liver, to release angiotensin I (Figure 12.1). The production of the decapeptide angiotensin I occurs in circulating blood, within blood vessel walls and within the nephron. Angiotensin converting enzyme (ACE), which is present in pulmonary circulation, kidney and vascular beds in several organs, converts angiotensin I to the octapeptide angiotensin II.

Angiotensin II enhances aldosterone secretion, increases glomerular filtration via its vasoconstrictive actions, indirect pressor effects on the central nervous system and increases motility of cardiac myocytes: the overall effects are to promote reabsorption of sodium and water in the proximal tubules and to stimulate the production of AVP.

12.1.3 *Aldosterone*

(See Chapter 8.) This mineralocorticoid is secreted in the zona glomerulosa of the adrenal cortex. Aldosterone stimulates sodium reabsorption in exchange for potassium and hydrogen ion secretion via cytoplasmic receptors – types I and II – in the renal collecting tubules. The type I receptors appear to be similar to the

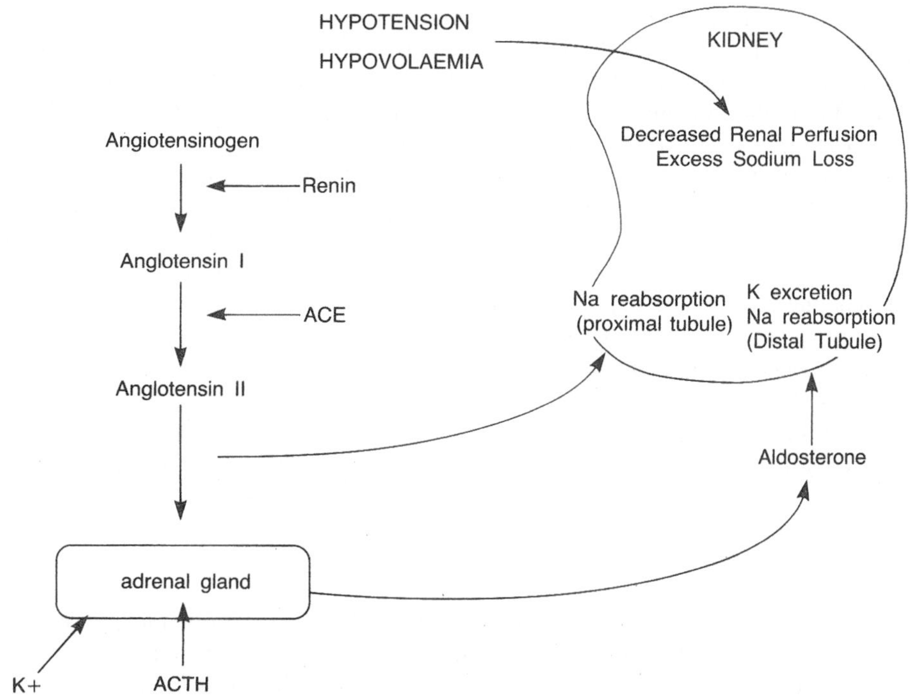

Figure 12.1 Renin–angiotensin–aldosterone axis system

hippocampal high-affinity glucocorticoid receptors with cortisol and corticosterone acting via both types of receptors to cause changes in mineralocorticoids (Funder, 1993). Several feedback mechanisms on aldosterone production exist, and these include plasma, adrenal and renal electrolyte concentrations and the renin–angiotensin axis (White, 1994).

12.1.4 *Atrial Natriuretic Factor (ANF or ANP)*

This polypeptide hormone is synthesized, stored and secreted by cardiac myocytes mainly of the right atrium. ANF is secreted in response to increased atrial pressure or stretch following excessive sodium intake or volume expansion. In the kidney, the multiple actions of this hormone can increase glomerular filtration rate, urinary fluid and electrolyte excretion, sodium secretion into the medullary collecting ducts, alter renal vascular resistance, and decrease inner medullary hypertonicity and sodium reabsorption by the tubular epithelial cells. ANF also alters the secretion of renin and aldosterone. The net result of these actions is to increase sodium excretion and urinary volume and lower urinary osmolality (de Wardener, 1982; Maack, 1988, 1992).

12.1.5 *Parathyroid Hormone (PTH or Parathormone) and Calcitonin*

These hormones and 25-dihydroxycholecalciferol (25-DHCC) regulate calcium homeostasis and therefore indirectly affect phosphate metabolism. PTH secretion is increased in response to hypocalcaemia whereas increased plasma calcium suppresses PTH secretion. PTH stimulates the renal production of 25-DHCC, bone formation and resorption and renal reabsorption of calcium (Kurokawa, 1994). The peptide hormone calcitonin acts in opposition to the effects produced by PTH (Paterson, 1974). Growth hormone, glucocorticoids, oestrogens, testosterone and the thyroid hormones also influence calcium metabolism.

Generally, all of these hormone measurements are used selectively in mechanistic studies, where the pharmacological or mode of action of the test compound suggests that the assays will be useful in determining the pathogenesis of renal or other changes, or to further characterize physiological reactions which are not desired or expected by design. Although all of these hormones have been assayed in various species, there are several factors including anaesthetics, stress, dietary intake of sodium and potassium, fluid intake, diurnal variation, and dark and light illumination cycles, which are known to affect these determinations, and therefore the analytical results require careful interpretation (Wells *et al.*, 1993).

12.2 Water, Sodium and Potassium Measurements

Water is lost via faeces, urine, salivation, insensible respiration and cutaneous water loss, with sensible perspiration of sweat occurring in some species. Vomitus and diarrhoea both affect fluid and electrolyte balance, and this balance is often altered where gastric irritation occurs in toxicological studies. Water excretion is governed by hypothalamic osmoreceptors which respond to raised osmolality and cause AVP secretion, followed by reabsorption of water in the renal collecting ducts: the second control mechanism operates via the action of aldosterone. Some examples of compounds which cause pharmacological modulation of electrolyte balance are given in Table 12.1 and further examples are given in this section (Velazquez and Wright, 1986; Stier and Itskovitz, 1986).

Overloading or water excess may occur when excessive thirst is induced by a test compound or where renal excretion does not act as a compensatory mechanism: this results in a lowering of the levels of plasma sodium, protein, albumin and osmolality. Fluid overloading may also occur when the rates for intravenous fluid administration are inappropriately high. The osmolality of a test solution for intravenous administration may affect fluid balance where the injection volumes are relatively large compared to the ECF volume of the animal (Michell, 1991).

Depletion of body water (or dehydration) may occur when food and water intakes are severely reduced, and values for plasma sodium and osmolality, total protein and albumin tend to increase. Packed cell volume (PCV or haematocrit, Hct) can be an important indicator of perturbations of fluid balance, but the relationship between PCV and plasma volume deficit is not linear (Boyd, 1981).

Osmolality is defined as the osmotic pressure exerted by a solution across a membrane, and it is proportional to the number of solute particles per unit weight of water and is expressed as mosmol/kg. Plasma and urine osmolality are commonly determined by methods using depression of freezing point; some laboratories use a vapour pressure method. An approximation of plasma osmolality can be obtained by adding the concentrations of glucose, urea and twice

Table 12.1 Some examples of pharmacological effects on electrolyte balance

Pharmacological action (examples of compounds)	Possible effect on plasma electrolyte
Effects on AVP	
Increased production of AVP (opiates, barbiturates, tolbutamide)	Hyponatraemia
Potentiation of AVP effect (chlorpropamide, acetoaminophen)	Hyponatraaemia
Suppression of AVP release (phenytoin, ethanol)	Hypernatraemia
Decreased effect of AVP on renal tubules and water diuresis (lithium)	Hypernatraemia
Diuretics	
Reduced bicarbonate reabsorption and increased potassium excretion in proximal tubules (carbonic anhydrase inhibitors)	Hypokalaemia Metabolic acidosis
Reduced reabsorption of sodium and chloride in distal tubules, increased excretion of potassium (thiazide diuretics)	Hyponatraemia Hypokalaemia Metabolic acidosis
Reduced reabsorption of sodium and chloride associated with diuresis (furosemide, loop diuretics)	Hyponatraemia Hypokalaemia
Reduced sodium reabsorption via aldosterone (spironolactone) blockage of sodium channels (potassium sparing diuretics)	Hyperkalaemia Hyponatraemia
Effects on renin–angiotensin axis	
Reduced production of angiotensin II (angiotensin converting enzyme inhibitors)	Hyperkalaemia Hyponatraemia
Competitive inhibition of renal angiotensin II receptors (angiotensin II antagonists)	Hyperkalaemia Hyponatraemia

the sodium concentration when all are expressed as mmol/L. Significant amounts of compounds with low molecular mass, e.g. ethanol and ethylene glycol can cause a marked discrepancy between measured and calculated osmolality values. Plasma osmolality measurements are of value when it is suspected that the calculated and measured osmolality values are markedly different or where some estimation of free water clearance or renal tubular function is required.

In the state of pure water deficiency, plasma osmolality increases; alterations in renal concentrating ability which result in water loss from the body are usually accompanied by loss of solute, but sometimes the water loss is excessive and leads to a hyperosmolar state. Urinary osmolality measurements can be used to monitor renal concentrating ability. The maximal urine concentrating and diluting ability varies widely between species, with rats and dogs able to produce more concentrated urine (higher than 2500 mosmols/kg body water) than found in man.

12.3 Electrolytes

Dietary electrolyte requirements vary, with some being required in small amounts or at rare intervals, while others such as potassium and phosphate must be ingested regularly to prevent deficiencies, as they are continually excreted. These dietary requirements also vary between species. Where the test compound or vehicle contains a significant amount of electrolyte, e.g. sodium or potassium salts, then results should be interpreted accordingly.

The interpretation of plasma electrolyte values may be further complicated where cardiac output directly alters fluid balance and indirectly affects other organ functions, e.g. kidney. As the kidneys receive approximately one-quarter of total cardiac output, renal toxicity affects electrolyte and fluid balance. Disturbances of the intra- and extracellular equilibrium of cations – sodium, potassium, calcium and magnesium – may occur with cardiotoxic compounds (Cobb and Michell, 1992).

12.3.1 *Sodium*

This is the most abundant cation in the body and its concentration is closely related to osmotic homeostasis, maintenance of body fluid volumes and neuromuscular excitability. Plasma sodium is freely filtered via the renal glomeruli, but about 70% is reabsorbed in the proximal tubules and 25% in the loop of Henle together with chloride ions and water. In the distal convoluted tubules, aldosterone modulates sodium reabsorption and sodium ions are exchanged for potassium and hydrogen ions. Several of the causes of hyper- and hyponatraemia are listed in Table 12.2 (Eger *et al.*, 1983; Crawford *et al.*, 1984; Hawks *et al.*, 1991).

12.3.2 *Potassium*

This cation is predominantly intracellular with plasma values approximately twentyfold less than the intracellular concentrations. The intracellular potassium gradient is maintained by an ATP-dependent active extrusion of sodium, which is balanced by the pumping of potassium and hydrogen ions into the cells (Wingo and Cain, 1993). The maintenance of potassium concentrations in the various body fluids is essential to maintain the membrane potentials required for neuromuscular excitability (see Chapter 10 – Cardiotoxicity and Myotoxicity). Several of the causes of hyper- and hypokalaemia are listed in Table 12.3. Potassium excess is rare and raised plasma potassium values are more often caused by artefacts.

12.3.3 *Chloride*

This is the major extracellular anion and it is important in the maintenance of electroneutrality and osmolality together with sodium. After filtration through the renal glomeruli, chloride is passively reabsorbed in the proximal convoluted tubules, actively reabsorbed in the loop of Henle by a 'chloride pump', and it is also reabsorbed with sodium in the distal tubules. Hypochloraemia is observed in

Table 12.2 Some causes of hyper- and hyponatraemia

Hypernatraemia

(1) With decreased body sodium
Extra-renal loss:
Diarrhoea, salivation, reduced ICF and ECF
Shock, oliguria, hypotension, reduced ECF

(2) With normal body sodium
Insufficient fluid intake or increased water loss

(3) With increased body sodium
Excessive sodium intake or reduced renal excretion

(Examples of compounds which raise plasma sodium include mineralo- and glucocorticoids)

Hyponatraemia

(1) With low ECF volume (hypovolaemia):
Extra-renal loss: vomiting, diarrhoea, inflammatory exudate
Renal loss: adrenal insufficiency, mineralocorticoid deficiency, osmotic diuresis, renal tubular acidosis

(2) With normal ECF volume (euvolaemia)
In acute or chronic water retention, inappropriate ADH secretion, glucocorticoid deficiency, chronic renal injury, hypothyroidism

(3) With increased ECF (hypervolaemia)
Renal failure, nephrotic syndrome, congestive cardiac failure, hypoproteinaemia

(4) Dietary insufficiency

(5) Artefactual
Pseudohyponatraemia
Hyper-osmolar hyponatraemia associated with hyperglycaemia following mannitol infusion

(Examples of compounds which lower plasma sodium include diuretics, angiotensin converting enzyme inhibitors and antagonists, cabamazepines and opiates)

metabolic acidosis, metabolic alkalosis, severe dehydration and gastrointestinal fluid loss. In most instances, plasma sodium and chloride values tend to parallel each other.

12.3.4 *Urinary Sodium, Potassium and Chloride*

The urinary concentrations of these cations are highly variable, being dependent on diet and subject to the effects of faecal contamination. Where urinary output is affected by dehydration or excessive fluid losses via the gastrointestinal tract, the data are also highly variable. Incorrectly low urinary electrolyte values may appear when water is accidentally added from water bottles to the urine samples by animal movements: such an event may sometimes be detected by measuring osmolality and noting unusually high urine volume together with a pale appearance of the sample. Perturbations of renal tubular function may be confirmed by urinary electrolyte measurements but given the various sites for electrolyte reabsorption, these may not necessarily identify the damaged region of the nephron.

Table 12.3 Some causes of hyper- and hypokalaemia

Hyperkalaemia

(1) Increased intake via infusion solutions or use of potassium salts

(2) Reduced potassium excretion due to:
  Hormonal effects – hypoaldosteronism, adrenal toxicity
  Renal effects – oliguria, chronic renal toxicity, diuresis

(3) Metabolic and respiratory acidosis

(4) Cellular necrosis, intravascular haemolysis

(5) Hypertonic states

(6) Artefactual
  Haemolysis
  Release from leucocytes and platelets
  Difficult blood collections
  Potassium containing anticoagulant, commonly EDTA

Hypokalaemia

(1) Decreased intake

(2) Extra-renal loss:
  Vomiting, diarrhoea, increased faecal loss
  (Examples of compounds which increase gastrointestinal loss include laxatives)

(3) Renal loss:
  Increased mineralocorticoids
  Hyperaldosteronism
  Renal tubular acidosis and chronic alkalosis
  Renal tubular injury
  (Examples of drugs which alter renal excretion include aminoglycosides, antineoplastic agents and diuretics)

(4) Redistribution of potassium between ECF and ICF – sometimes associated with alkalosis
  (Examples of compounds with these effects include β-adrenergic agonists such as salbutamol, insulin and folate)

(5) Artefactual
  In hyperlipidaemia and hyperproteinaemia

12.3.5 *Calcium and Inorganic Phosphate*

Approximately 99% of the body's calcium is contained in bone as the hydroxyapatite, and calcium is the most abundant mineral in the body. This cation is involved in neuromuscular transmission, cardiac and skeletal muscular contraction and relaxation, bone formation, coagulation, cell growth, membrane transport mechanisms and enzymatic reactions. The hormonal controls for body calcium were briefly described in an earlier section (see also DeLuca, 1988).

Approximately 40 to 50% of the plasma calcium is free or ionized, with the remaining plasma fraction bound to plasma proteins, mainly albumin. Acidosis increases plasma ionized calcium concentrations, whereas alkalosis causes a decrease due to the effects of pH in the ECF or on protein binding. Failure to consider any changes in plasma proteins which may be occurring simultaneously,

Table 12.4 Some causes of hyper- and hypocalcaemia

Hypercalcaemia	
(1)	Hyperparathyroidism
(2)	Malignancy
(3)	Other endocrine disorders Hyperthyroidism Hypoadrenalism
(4)	Dehydration
(5)	Hyperproteinaemia particularly hyperalbuminaemia
(6)	Artefactual Poor venepuncture technique
	(Examples of drugs which raise plasma calcium include thiazides, lithium, calciferol containing rodenticides)
Hypocalcaemia	
(1)	Renal injury associated with impaired hydroxylation of 25-hydroxycholecalciferol or acidosis
(2)	Inadequate nutritional intake of calcium, vitamin D or both
(3)	Hypoparathyroidism
(4)	Acute pancreatitis
(5)	Hypoproteinaemia particularly hypoalbuminaemia
	(Examples of compounds which lower plasma calcium include calcitonin, diuretics, anticonvulsants, fluoride and ethylene glycol)

is a major pitfall in the interpretation of plasma calcium measurements. A shift of albumin related to globulin may also modify the amounts of calcium carried by these protein fractions.

Plasma inorganic phosphate (sometimes referred to as inorganic phosphorus) metabolism is governed by the same hormones as calcium, but inorganic phosphate levels are more sensitive than calcium to dietary intake and renal excretion. Minor changes of plasma phosphate may be observed in renal, pre-renal and post-renal azotaemia (Kempson and Dousa, 1986). Both plasma inorganic phosphate and calcium levels vary with age, being higher in very young animals (Kiebzak and Sacktor, 1986). Tables 12.4 and 12.5 show some of the causes for changes in plasma calcium and inorganic phosphate values.

12.3.6 Magnesium

This cation is essential for many enzyme reactions, neuromuscular activity and bone formation; it is the second most abundant intracellular cation. When compared with calcium, much less is known about homeostatic mechanisms for magnesium, although the normally narrow ranges for plasma magnesium imply close homeostatic control. Table 12.6 shows the conditions where alterations of plasma magnesium occur. This cation is less commonly measured in toxicological studies and indeed in human clinical medicine, but there is a growing interest in

Table 12.5 Some causes of hyper- and hypophosphataemia

Hyperphosphataemia
(1) Increased dietary intake
(2) Tissue necrosis
(3) Acidosis
(4) Altered bone metabolism
(5) Decreased renal excretion
(6) Artefactual – delayed sample separation or haemolysis
(Examples of compounds which raise plasma inorganic phosphate include anabolic steroids, furosemide and thiazides)

Hypophosphataemia
(1) Redistribution of phosphate between ECF and ICF
(2) Decreased nutritional intake
(3) Hepatic disease
(4) Hyperparathyroidism
(5) Infection
(6) Increased excretion
(Examples of compounds which lower plasma inorganic phosphate include antacids, renal tubular toxins, ethanol, intravenous glucose and insulin)

its measurement with compounds such as ciclosporin (Borland *et al.*, 1985; Evans, 1988) and cisplatin (Magil *et al.*, 1986).

12.3.7 Electrolyte Measurements

Most laboratories measure sodium or potassium either by flame photometry or ion-selective electrodes: more recent methods involving chromogenic or enzymatic procedures for these cations have not been widely used for animal samples. Ion-selective electrodes or colorimetric methods are used for chloride determinations but halogenated test compounds may cause technical interferences in these assays. Colorimetric methods are generally used for plasma total calcium and magnesium measurements, and some investigators have used ion-selective electrodes to measure plasma ionized calcium.

The ranges for electrolytes vary between species and these ranges are partly dependent on the site and method for blood collection (Neptun *et al.*, 1985; Matsuzawa *et al.*, 1994; see Chapter 3). Use of plasma rather than serum reduces the times required for the blood separation procedures, but may result in minor differences between serum and plasma. Plasma potassium (and inorganic phosphate) levels rise if the separation of whole blood is delayed: this effect on potassium levels is also seen in haemolysed samples, although this is less marked in the dog where the erythrocytic concentrations of potassium are lower. Tissue injury which can occur during blood collection procedures can lead to falsely high potassium values.

Lithium heparinate is the anticoagulant of choice for plasma electrolyte

Table 12.6 Some causes of hyper- and hypomagnesaemia

Hypermagnesaemia
(1) Decreased excretion in acute and chronic renal failure
(2) Increased intake e.g. magnesium salts of test compounds
(3) Cellular necrosis
(4) Adrenocortical hypofunction

Hypomagnesaemia
(1) Reduced nutritional intake
(2) Malabsorption
(3) Increased gastrointestinal fluid loss
(4) Pancreatitis
(5) Increased renal excretion
(6) Hypoparathyroidism
(7) Hyperthyroidism
(8) Hypocalcaemia
(9) Hypokalaemia
(Examples of compounds which lower plasma magnesium include aminoglycosides, cisplatin, ciclosporin and loop diuretics)

measurements (except of course for plasma lithium). Falsely elevated potassium values result when sequestrenated (EDTA) samples are measured, and the anticoagulant sodium heparinate will give incorrectly elevated sodium values. Hyperlipidaemia and hyperproteinaemia may give rise to artefactual reductions of plasma sodium (as measured by indirect ion-selective electrodes or flame photometry), i.e. pseudohyponatraemia.

Total plasma (or urine) calcium and magnesium levels can be measured by several colorimetric methods but relatively large fractions of these two cations are protein bound, i.e. more than 30% of these total plasma ions concentrations. Ionized fractions are more difficult to measure and require careful blood collection procedures. Where marked alterations of plasma proteins concentrations occur, particularly albumin, it is necessary to consider the effects on the protein-bound divalent ions.

Any extreme change in the ionic balance of one cation can influence the concentrations of other cations and anions. Cationic changes usually are accompanied by perturbations of anionic concentration and fluid balance. The 'anion gap' is customarily calculated by subtracting the sum of chloride and bicarbonate concentrations from the sum of the sodium and potassium concentrations in either plasma or urine: the difference reflects the so-called unmeasured anions such as plasma proteins. For plasma, the anion gap is approximately 10 to 20 mmol/L, and this calculation can be used to check abnormal values. In the absence of a measured plasma bicarbonate value but with a knowledge of both plasma glucose and urea values it is possible to estimate the plasma bicarbonate value using this formula, and use this as an indication of metabolic acidosis associated with renal or gastrointestinal loss of bicarbonate (Feldman and Rosenberg, 1981). Unless plasma samples are measured promptly following collection, there is little value in measuring plasma bicarbonate because of the loss of volatile carbon dioxide on exposure to air.

12.4 Acid–Base Balance

The hydrogen ion concentration (or pH) of the ECF is maintained within narrow limits, with perturbations affecting metabolism and molecular structures, e.g. proteins. Severe alterations of electrolyte concentrations are often accompanied by changes in acid–base balance with regulatory mechanisms involving extra- and intracellular buffer systems, alveolar ventilation, expiry of carbon dioxide and renal excretion of hydrogen ion. The causes for perturbations may be respiratory, metabolic or of mixed origin; the effects may be of variable duration and severity within treatment groups with differing compensatory and non-compensatory mechanisms. Measurement of oxygen (pO_2) can be important when testing compounds which bind with haemoglobin and therefore cause a shift in blood oxygenation.

Remember that acid–base balance can be affected by the use of anaesthesia and sedatives: it is partly because of these effects that acid–base balance measurements are not included routinely in toxicological studies, although they may be used in pharmacological or inhalation studies. These measurements require carefully controlled conditions if they are to be meaningful and samples need to be analysed promptly. Published data are available for larger laboratory animals, e.g. dog (Cornelius and Rawlings, 1981), although there is paucity of data for small animals: different methods of sampling produce variations in results (Upton and Morgan, 1975).

References

Borland, I. A., Gosney, J. R., Hillis, A. N., Williamson, E. P. M. & Sells, R. A. (1985) Hypomagnesaemia and cyclosporin toxicity. *Lancet*, **i**, 103–4.

Boyd, J. W. (1981) The relationships between blood haemoglobin concentrations, packed cell volume and plasma protein concentration in dehydration. *British Veterinary Journal*, **137**, 166–72.

Cobb, M. & Michell, A. R. (1992) Plasma electrolyte concentrations in dogs receiving diuretic therapy for cardiac failure. *Journal of Small Animal Practice*, **33**, 526–9.

Cornelius, I. M. & Rawlings, C. A. (1981) Arterial blood gas and acid–base values in dogs with various diseases and signs of disease. *Journal of the Veterinary Medical Association*, **178**, 992–5.

Crawford, M. A., Kittelson, M. D. & Fink, G. D. (1984) Hypernatraemia and adipsia in a dog. *Journal of American Veterinary Medical Association*, **184**, 818–21.

DeLuca, H. F. (1988) The vitamin D story – a collaborative effort of basic science and clinical medicine. *FASEB Journal*, **2**, 224–36.

Eger, C. E., Robinson, W. F. & Huxtable, C. R. R. (1983) Primary aldosteronism (Conn's syndromes) in a cat: a case report and review of comparative aspects. *Journal of Small Animal Practice*, **24**, 293–307.

Evans, G. O. (1988) Hypomagnesaemia, hypoalbuminaemia and plasma lipid changes in rats following the oral administration of ciclosporin. *Comparative Biochemistry and Haematology*, **89C**, 375–6.

FELDMAN, B. F. & ROSENBERG, D. P. (1981) Clinical use of anion and osmolal gaps in veterinary medicine. *Journal of American Veterinary Medical Association*, **178**, 396–8.

FUNDER, J. W. (1993) Aldosterone action. *Annual Review of Physiology*, **55**, 115–30.

HAWKS, D., GIGER, U., MISELIS, R. & BOVEE, K. C. (1991) Essential hypernatraemia in a young dog. *Journal of Small Animal Practice*, **32**, 420–4.

HORN, A. M., ROBINSON, I. C. A. F. & FINK, G. (1985) Oxytocin and vasopressin in rat hypophyseal portal bloods: experimental studies in normal and Brattleboro rats. *Journal of Endocrinology*, **104**, 221–4.

KEMPSON, S. A. & DOUSA, T. P. (1986) Current concepts on regulation of phosphate transport in renal proximal tubules. *Biochemical Pharmacology*, **35**, 721–6.

KIEBZAK, G. M. & SACKTOR, B. (1986) Effect of age on renal conservation of phosphate in the rat. *American Journal of Physiology*, **251**, F399–407.

KUROKAWA, K. (1994) The kidney and calcium homeostasis. *Kidney International*, **45**, suppl. 44, S97–105.

LARAGH, J. H. (1985) Atrial natriuretic hormone, the renin–aldosterone axis and blood pressure – electrolyte homeostasis. *New England Journal of Medicine*, **313**, 1330–40.

MAACK, T. (1988) Functional properties of atrial natriuretic peptide and its receptors. In Davidson, A. M. (ed.), *Nephrology*, Vol. 1. pp. 123–60. London: Baillière Tindall.

(1992) Receptors of atrial natriuretic factor. *Annual Review of Physiology*, **54**, 11–27.

MAGIL, A. B., MAVICHAK, V., WONG, N. L. M., QUAMME, G. A., DIRKS, J. H. & SURTON, R. A. L. (1986). Long-term morphological and biochemical observations in Cisplatin induced hypomagnesaemia in rats. *Nephron*, **43**, 223–30.

MATSUZAWA, T., TABATA, H., SAKAZUME, M., YOSHIDA, S. & NAKAMURA, S. (1994) Comparison of the effect of bleeding site on haematological and plasma chemistry values of F344 rats: the inferior vena cava, abdominal aorta and orbital venous plexus. *Comparative Haematology International*, **4**, 207–11.

MICHELL, A. R. (1991) Regulation of salt and water balance. *Journal of Small Animal Practice*, **32**, 135–45.

MOHR, E. & RICHTER, D. (1994) Vasopressin in the regulation of body fluids. *Journal of Hypertension*, **12**, 345–8.

NEPTUN, D. A., SMITH, C. N. & IRONS, R. D. (1985) Effect of sampling site and collection method on variations in baseline clinical pathology parameters in Fischer-344 rats. I. Clinical Chemistry. *Fundamental and Applied Toxicology*, **5**, 1180–5.

OPARIL, S. & HABER, E. (1974a) The renin–angiotensin system. Part 1. *New England Journal of Medicine*, **290**, 389–401.

(1974b) The renin–angiotensin system. Part 2. *New England Journal of Medicine*, **290**, 446–57.

PATERSON, C. R. (1974) *Metabolic Disorders of Bone*. Oxford: Blackwell Scientific.

ROBERTSON, G. L. (1987) Physiology of the ADH secretion. *Kidney International*, **32**, Suppl. 21, 20–61.

STIER, T. & ITSKOVITZ, H. D. (1986) Renal calcium metabolism and diuretics. *Annual Review of Pharmacology and Toxicology*, **26**, 101–16.

UPTON, P. K. & MORGAN, D. J. (1975) The effect of sampling technique on some blood parameters in the rat. *Laboratory Animals*, **9**, 85–91.

VELAZQUEZ, H. & WRIGHT, F. S. (1986) Control by drugs of renal potassium handling. *Annual Review of Pharmacology and Toxicology*, **26**, 293–309.

WARDENER, H. E. DE (1982) The natriuretic hormone. Recent developments. *Clinical Science and Molecular Medicine*, **63**, 415–20.

WELLS, T., WINDLE, R. J., PEYSNER, K. & FORSLING, M. L. (1993) Inter-colony variation in fluid balance and its relationship to vasopressin secretion in male Sprague–Dawley rats. *Laboratory Animals*, **27**, 40–6.

WHITE, P. C. (1994) Disorders of aldosterone biosynthesis and action. *New England Journal of Medicine*, **331**, 250–8.

WINGO, C. & CAIN, B. D. (1993) The renal H-K-ATPase. Physiological significance and role in potassium homeostasis. *Annual Review of Physiology*, **55**, 323–47.

General References

BECK, L. H. (ed.) (1981) Symposium on body fluid and electrolyte disorders, *Medical Clinics of North America*.

ELIN, R. J. (1987) Assessment of magnesium status. *Clinical Chemistry*, **33**, 1965–70.

JAMIESON, M. J. (1985) Hyponatraemia. *British Medical Journal*, **290**, 1723–8.

MAXWELL, H. H., KLEEMAN, C. R. & NARINS, R. G. (eds) (1987) *Clinical Disorders of Electrolyte Metabolism*, 4th Edn. New York: McGraw-Hill.

ROBINSON, A. G. (1985) Disorders of anti-diuretic hormone secretion. *Clinics in Endocrinology and Metabolism*, **14**, 55–89.

ROSE, B. D. (1984) *Clinical Physiology of Acid–Base and Electrolyte Disorders*, New York: McGraw-Hill.

WILLARD, M. D. (1989) Electrolyte and acid–base abnormalities. In Willard, M. D., Tvedten, H. & Turnwald, G. H. (eds), *Small Animal Clinical Diagnosis by Laboratory Methods*, pp. 103–20. Philadelphia: W. B. Saunders.

WINTERS, R. W. (1966) Terminology of acid–base disorders. *Annals of the New York Academy of Science*, **133**, 211–14.

13

Lipids

G. O. EVANS

Lipids having poor solubility in water are transported *in vivo* by the various lipoprotein fractions where they are complexed with peptides. The lipids are carried in hydrophilic forms consisting of an outer layer of protein (apolipoprotein) and polar lipids (phospholipid and unesterified cholesterol) around an inner hydrophobic core of the neutral lipids – triglyceride and cholesterol esters. These core sizes change according to the neutral lipid content.

The lipoproteins vary in size, density, lipid composition and the apolipoproteins (the protein fractions), and they can be classified by either the flotation rate determined by ultracentrifugation or by their electrophoretic mobilities. In descending order of size, the broad lipoprotein fractions are:

- Chylomicrons (or chylomicra)
- VLDL Very low density lipoproteins (pre-betalipoproteins)
- IDL Intermediate density lipoproteins (slow pre-beta-lipoproteins)
- LDL Low density lipoproteins (beta-lipoproteins)
- HDL High density lipoproteins (alpha-lipoproteins)

The apolipoproteins are labelled alphabetically from A to E with additional subsets denoted by roman numerals, e.g. apolipoprotein A-II.

The chylomicrons mainly consist of triglycerides with small amounts of phospholipids, esterified and free cholesterol and apolipoproteins. The VLDL particles contain proportionally more protein and cholesterol but less triglycerides than the chylomicrons. The LDL fraction contains more protein but less triglyceride than the chylomicrons and the VLDL. The HDL fraction contains the lowest proportion of triglyceride but it has the highest protein content of the lipoprotein fractions.

Following absorption through the intestinal wall, cholesterol and triglycerides are converted to the triglyceride-rich chylomicrons which are then transported to the adipose tissue and muscle (Figure 13.1) where triglycerides are hydrolysed by the lipoprotein lipases. Chylomicron 'remnants' are transported to hepatic receptor sites. Whereas triglycerides serve mainly as energy carriers and storage substances,

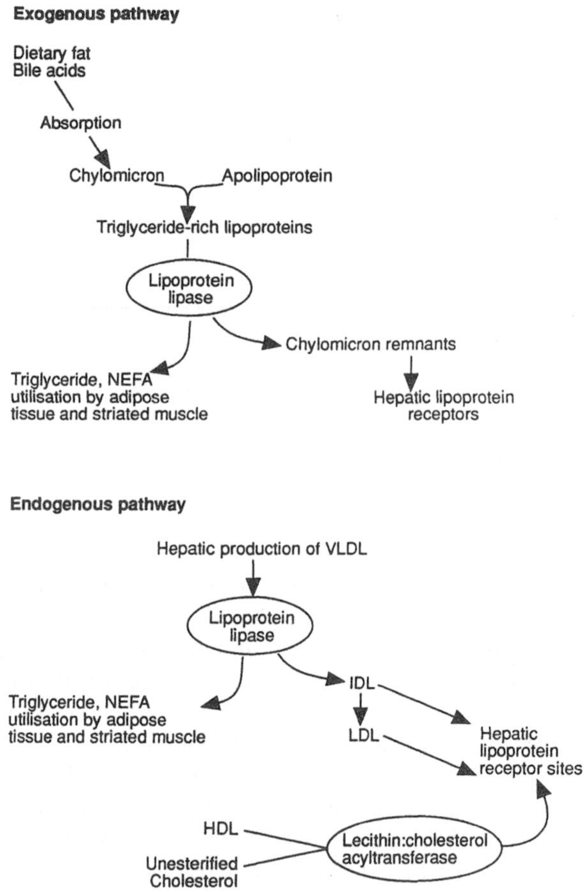

Figure 13.1 Simplified pathways for lipid metabolism

cholesterol is required for the formation of cell membranes and as the precursor for steroid hormone synthesis. Cholesterol is also excreted from the liver as bile acids. Most of the requirements for cholesterol are met by endogenous synthesis mainly in the liver, with the exogenous supplementation from the diet.

The liver synthesizes VLDL, IDL and LDL, and the LDL are then transported to non-hepatic tissues to undergo further metabolism. HDL may be released directly from the liver as 'nascent HDL', or may be formed from HDL precursors from the chylomicrons in the small intestine or hepatic VLDL fractions. Although the liver is the main site of uptake for chylomicrons, the perisinusoidal bone marrow macrophages also account for a significant proportion of the total uptake of chylomicrons in several species (Ross, 1986; Hussain *et al.*, 1989). (For reviews of lipid metabolism, see Levy, 1981; Brown and Goldstein, 1987; Gotto, 1988; Kreuzer and von Hodenberg, 1994; Watson and Barrie, 1993).

There are marked differences between species for plasma cholesterol, triglycerides and lipoproteins (Alexander and Day, 1973; Chapman, 1980; Lehmann *et al.*, 1993). The proportions of protein, cholesterol and triglycerides vary in the lipoproteins of the different species (Table 13.1). Whereas LDL is the

Table 13.1 Relative plasma lipid levels for common laboratory animals, values increase downwards with highest at the bottom

Cholesterol	Triglycerides	HDL
Guinea pig	Guinea pig	Guinea pig
Rat	Dog	Rat
Rabbit	Rat	Rabbit
Mouse	Hamster	Hamster
Hamster	Monkey	Monkey
Dog	Rabbit	Mouse
Monkey	Mouse	Dog

dominant plasma lipoprotein fraction in the non-human primates, dogs and hamsters, the HDL fraction is the dominant fraction in the rat. In contrast to other herbivores, 80% of plasma cholesterol in guinea pigs is transported as LDL. Structural differences of the apolipoproteins are also found, but there are close similarities between some species; for example several of the apolipoproteins in the common marmoset and man are similar (Crook *et al.*, 1990).

Variations in plasma cholesterol associated with age occur in several species. In the rat, both plasma cholesterol and triglyceride increase during the first nine months of life (Carlson *et al.*, 1968), while in Watanabe rabbits both serum cholesterol and triglycerides decrease from week 10 during the first year of life (Lind *et al.*, 1990). Breed type, age and gender in dogs appear to have no effect on cholesterol and lipoprotein concentrations (Barrie *et al.*, 1993).

13.1 Animal Models of Hyperlipidaemia

The association between the development of atherosclerosis in man and changes of plasma lipoproteins has led to the development and use of several animal models to help our understanding of familial hypercholesterolaemia (Beynen, 1988; Overturf and Loose-Mitchell, 1992). Examples include Yoshida and Zucker rats and the heritable hyperlipidaemic Watanabe rabbits (Watanabe, 1980).

Dietary manipulation can alter plasma cholesterol and triglycerides to varying degrees and nutritional effects have been described in several species including rats (Mahley and Holcombe, 1977; Van Zutphen and den Bieman, 1981), hamsters (Weingand and Daggy, 1991), gerbils (Temmerman *et al.*, 1989), and marmosets (McIntosh *et al.*, 1984).

13.2 Toxicological Studies

Sometimes relatively small changes of plasma cholesterol (and triglycerides) are found in toxicological studies, but the reasons for these changes are not always identifiable. From our knowledge of lipid metabolism, the possible reasons for toxicological effects include nutritional, intestinal, hepatic, hormonal, cardiac and renal factors. Reduced food and energy intake resulting from toxicity will cause the body to release energy from the adipose and muscle tissues to compensate for these deficiencies, and these changes may be reflected by the plasma triglyceride and cholesterol levels.

With hepatotoxicity, there may be several associated perturbations of lipoprotein metabolism. In acute hepatic injury, levels of hepatic enzymes may be reduced, e.g. triglyceride lipase, thus elevating plasma triglycerides with a concomitant decrease in cholesterol. Prolonged cholestasis will result in bile not reaching the duodenum with consequential malabsorption of fat and fat-soluble vitamins. As the plasma bile acids increase in cholestasis, then plasma cholesterol also tends to rise.

Impaired hepatic protein metabolism will also affect lipoprotein synthesis resulting in hypolipoproteinaemia and hypotriglyceridaemia. Hypocholesterolaemia has been described in rodent studies for several peroxisome proliferators (Reddy and Lalwani, 1987).

Changes in plasma cholesterol may occur secondarily to hormonal changes, e.g. during pregnancy, in diabetes mellitus and alterations of corticosteroid metabolism. Alternatively gross changes in lipid metabolism may affect hormonal patterns particularly where cholesterol is used for hormone synthesis, i.e. steroid hormones. In thyroid disease or toxic injury, plasma cholesterol levels are reduced in hyperthyroidism and conversely increased in hypothyroidism. Nephrotic syndrome is characterized by the hyperlipidaemia, hypoproteinemia and hyperproteinuria in several species, and lipid changes may also be observed with chronic renal damage.

In testing novel pharmaceuticals which are aimed at modulating plasma cholesterol and lipoprotein fractions (Shepherd and Packard, 1988), various effects may be observed in safety evaluation studies of these compounds. The hepatic enzymes governing lipid metabolism are the target for several compounds used for the treatment of hyperlipidemia. These compounds include fibric acid derivatives (Zimetbaum *et al.*, 1991) and HMG-CoA reductase inhibitors (Frishman *et al.*, 1989). Several anti-hypertensive agents, including non-specific or selective beta-adrenergic blocking drugs, are also known to affect plasma lipid profiles (Chait, 1989; Karge *et al.*, 1989).

13.3 Measurements of Plasma Lipids

13.3.1 Blood Samples

Given the effect of diet of plasma lipids, blood samples should be collected after a period of fasting appropriate for the species. Although heparinized plasma (or serum) lipoproteins are relatively stable, some investigators use sequestrenated (EDTA) samples as this chelates heavy metals, which promote auto-oxidation of unsaturated fatty acids and cholesterol, and it inhibits phospholipase-c activity due to bacterial contamination. Lipoprotein measurements should be performed promptly with fresh samples to avoid problems due to lipoprotein instability during storage.

The simple observation of a creamy layer or dispersed opalescence in the plasma indicating hyperchylomicronaemia (associated with high triglyceride levels) can be important. This lipid layering may be encouraged by storing the samples overnight at 4°C. Grossly lipaemic samples can cause several analytical problems if these samples are not homogeneous at the time of analysis. Lipaemia can interfere with several routine photometric analyses, and enhance the effects of interference due to haemolysis (Brady and O'Leary, 1994; Sampson *et al.*, 1994). Some reagents contain solubilizers which act on the lipid micelles and reduce interference effects due to lipaemia (see Chapter 3).

13.3.2 *Analytical Methods*

Many methods have been used for the analysis of the various lipid and lipoprotein fractions, and these are considered under the following headings:

- Total and free cholesterol
- Triglycerides
- Lipoproteins
- Other lipids assays including phospholipids, non-esterified fatty acids (NEFA), fatty acids and total lipids.

Total and Free Cholesterol

Enzymatic methods are commonly used for this measurement and many of the current methods use a cholesterol esterase to hydrolyse lipoprotein–cholesterol esters, followed by further reactions in which cholesterol oxidase links with a peroxidase–chromogen system (Richmond, 1992). The esterases used in these analytical systems show varying degrees of specificity towards the different cholesterol esters, and problems have occurred where these reagents are used for several species: for example, in rats where there is a fivefold higher concentration of cholesteryl arachidonate ester compared with human plasma, plasma cholesterol may be underestimated with some reagents (Demacker *et al.*, 1983; Noel *et al.*, 1983; Wiebe and Bernert, 1984; Evans, 1986).

Triglycerides

These can be measured with one of several enzymatic reagents (Klotzsch and McNamara, 1990). Again the effectiveness of some of the enzyme reagents appears to differ with the various species, e.g. underestimation of triglycerides in porcine sera using the lipase–glycerol method (Tuten *et al.*, 1993). Some methods also measure plasma free glycerol together with triglycerides or alternatively use a secondary measurement to eliminate the possible differences due to the free glycerol. The contribution of free glycerol to total triglycerides values varies from species to species (Weingand and Hudson, 1989). Ideally, a free glycerol blank should be carried out without glycerol kinase and/or with lipase containing reagents without one of the final chromogen system, but in most instances it is not necessary to alter the common triglyceride methods unless a more detailed explanation of these results is required.

Lipoproteins

While ultracentrifugation remains the reference method for studying lipoproteins (Hollanders *et al.*, 1986), few toxicology laboratories have direct access to this technology. Other alternative techniques include affinity chromatography and selective precipitation for HDL or LDL. The problems associated with precipitation methods for human plasma (Warnick and Albers, 1978; Hoffman *et al.*, 1985) are increased with plasma samples from laboratory animals. The differing proportions of lipoproteins in the plasmas of the various species require careful selection of analytical conditions with these precipitation techniques, if meaningful results are to be obtained (Sjoblom and Eklund, 1989; Tschantz and Sunahara, 1993). Selective precipitation techniques for HDL and LDL can be used, but the efficiency and selectivity of the precipitation techniques should be checked by alternative techniques such as electrophoresis or ultracentrifugation. Indiscriminant use of reagents optimized for use with human plasma may yield erroneous results when used with other species. Use of immunoprecipitation reagents for lipoprotein fractions also needs to be carefully evaluated for specificity with laboratory animal samples.

The Friedewald formula can be used to determine LDL-cholesterol with human samples by measuring plasma cholesterol, triglycerides and HDL, and substituting the values in the following:

$$\text{LDL cholesterol} = \text{Total cholesterol} - \text{HDL cholesterol} - \frac{\text{Triglyceride}}{219}$$

where values are expressed as mmol.l^{-1} and where total ^{219}triglyceride is $<4.5\,\text{mmol.l}^{-1}$ (Friedewald *et al.*, 1972; Rifai *et al.*, 1992). Values for LDL-cholesterol cannot be calculated with this formula for laboratory animals where the proportions of the lipoprotein fractions are markedly different.

Lipoprotein electrophoresis A relatively simple way to support the findings from plasma cholesterol and triglycerides is to use an electrophoretic method for separating the major lipoprotein fractions. Various support media, buffers systems and electrophoretic equipment and detection reagents have been employed (Groulade *et al.*, 1981; Oppermann *et al.*, 1983).

Other Lipid Assays

For the majority of toxicology studies, measurements of plasma cholesterol and triglycerides are sufficient. However, some laboratories also include measurements of total lipids, non-esterified fatty acids (NEFA) or phospholipids. For apolipoproteins, immunoelectrophoretic and nephelometric assays have been performed in several species (Chirtel *et al.*, 1989) but these are also not commonly used in toxicology studies.

References

Alexander, C. & Day, C. E. (1973) Distribution of serum lipoproteins of selected vertebrates. *Comparative Biochemistry and Physiology*, **46B**, 295–312.

Barrie, J., Watson, T. G., Stear, M. J. & Nash, A. S. (1993) Plasma cholesterol and lipoprotein concentrations in the dog: the effects of age, gender and endocrine disease. *Journal of Small Animal Practice*, **34**, 507–12.

Beynen, A. C. (1988) Animal models in cholesterol metabolism studies. In Beynon, A. C. & Solleveld, H. A. (eds), *New Developments in Biosciences: Their Implications for Laboratory Animal Welfare*, pp. 279–88. The Netherlands: Martinus Nijhoff.

Brady, J. & O'Leary, N. (1994) Interference due to lipaemia in routine photometric analysis – survey of an underrated problem. *Annals of Clinical Biochemistry*, **31**, 281–8.

Brown, M. S. & Goldstein, J. L. (1987) The hyperlipoproteinaemias. In Braunwald, E. (ed.), *Harrison's Principles of Internal Medicine*, 11th Edn, pp. 1650–61. New York: McGraw-Hill.

Carlson, L. A., Froberg, S. O. & Nye, E. R. (1968) Effect of age on blood and tissue lipids in the male rat. *Gerontologia*, **14**, 65–79.

Chait, A. (1989) Effects of antihypertensive agents on serum lipids and lipoproteins. *American Journal of Medicine*, **86**, Suppl. 1B, 5–7.

Chapman, M. J. (1980) Animal lipoproteins: chemistry, structure and comparative aspects. *Journal of Lipid Research*, **21**, 789–848.

Chirtel, S. J., Coutlakis, P. J., Chambers, L. L. & Lakshman, M. R. (1989) A novel use of endpoint nephelometry to standardise the rate nephelometric assay of human and rat plasma apoprotein AB_1. *Journal of Laboratory and Clinical Medicine*, **113**, 632–41.

Crook, D., Weisgraber, K. H., Rall, S. C. & Mahley, R. W. (1990) Isolation and characterisation of several plasma lipoproteins of common marmoset monkey. *Arteriosclerosis*, **10**, 625–32.

Demacker, P. N. M., Boerma, G. J. M., Baadenenhuij, H., van Strik, R., Leijinse, B. & Jansen, A. P. (1983) Evaluation of accuracy of 20 different test kits for the enzymic determination of cholesterol. *Clinical Chemistry*, **29**, 1916–22.

Evans, G. O. (1986) The use of three esterase kits to measure plasma cholesterol concentrations in the rat and three other species. *Journal of Comparative Pathology*, **96**, 551–6.

Friedewald, W. T., Levy, R. J. & Fredrickson, D. S. (1972) Estimation of the concentration of low-density lipoprotein cholesterol in plasma without the use of the preparative ultracentrifuge. *Clinical Chemistry*, **18**, 499–509.

Frishman, W. H., Zimetbaum, P. & Nadelmann, J. (1989) Lovastatin and other HMG-CoA reductase inhibitors. *Journal of Clinical Pharmacology*, **29**, 975–82.

Gotto, A. M. (1988) Lipoprotein metabolism and the etiology of hyperlipidaemia. *Hospital Practice*, **23**, 4–13.

Groulade, P., Groslambert, P., Foulon, T. & Groulade, J. (1981) Electrophorèse des lipoproteines seriques de l'homme et du chien adultes normaux. *Bulletin Academy Nationale Médicine*, **165**, 1243–50.

Hoffman, G. E., Hiefinger, R., Weiss, L. & Poppe, W. (1985) Five methods for measuring low-density lipoprotein cholesterol concentration in serum compared. *Clinical Chemistry*, **31**, 1729–30.

Hollanders, B., Mougin, A., N'Diaye, F., Hentz, E., Aude, X. & Girard, A. (1986) Comparison of the lipoprotein profiles obtained from rat, bovine, horse, dog, rabbit and pig serum by a new two-step ultracentrifugal gradient procedure. *Comparative Biochemistry and Physiology*, **84B**, 83–9.

Hussain, M. M., Mahley, R. W., Boyles, J. R., Lindquist, P. A., Brecht, W. J. & Innerarity, T. L. (1989) Chylomicron metabolism. Chylomicron uptake by bone marrow in different animal species. *Journal of Biological Chemistry*, **264**, 17931–8.

Karge, W. H., Weiner, E. J., Kowala, M. C., Graham, R. M. & Nicolosi, R. J. (1989) Effects of Prazosin on hemodynamics, hemostasis, and serum lipid and lipoprotein levels in normal and hypercholesterolemic monkeys. *American Journal of Medicine*, **86**, Suppl. 1B, 19–23.

Klotzsch, S. G. & McNamara, J. R. (1990) Triglyceride measurements: a review of methods and interferences. *Clinical Chemistry*, **36**, 1605–13.

Kreuzer, J. & von Hodenberg, E. (1994) The role of apolipoproteins in lipid metabolism and atherogenesis: aspects in man and mice. *Journal of Hypertension*, **12**, 113–18.

Lehmann, R., Bhargava, A. S. & Gunzel, P. (1993) Serum lipoprotein patterns in rats, dogs and monkeys including method comparison and influence of menstrual cycle in monkeys. *European Journal of Clinical Chemistry and Clinical Biochemistry*, **31**, 633–7.

Levy, R. I. (1981) Cholesterol, lipoproteins, apoproteins, and heart disease: present states and future prospects. *Clinical Chemistry*, **27**, 653–62.

Lind, B. M., Littbarski, R., Hohlbach, G. & Moller, K. O. (1990) Long term investigations of serum cholesterol, serum triglyceride and HDL cholesterol in heritable hyperlipidemic rabbits. *Zeitschrift Versuchstierkunde*, **33**, 245–9.

Mahley, R. W. & Holcombe, K. S. (1977) Alterations of the plasma lipoproteins and apolipoproteins following cholesterol feeding in rats. *Journal of Lipid Research*, **18**, 314–24.

McIntosh, G. H., Bulman, F. H., Illman, R. J. & Topping, D. L. (1984) The influence of age, dietary cholesterol and vitamin C deficiency on plasma cholesterol concentration in the marmoset. *Nutritional Reports International*, **29**, 673–82.

Noel, A.-P., Dupras, R. & Filion, A.-M. (1983) The activity of cholesteryl ester hydrolysis in the enzymatic determination of cholesterol: comparison of five enzymes obtained commercially. *Clinical Chemistry*, **129**, 464–71.

Oppermann, P., Hubner, G. & Ehlers, D. (1983) Zur Trennung der Lipoproteine des Rattenserums mit Hilfe der Dsikelektrophorese. *Zeitschrift für Medezin, Laboratorium Diagnostik*, **24**, 200–3.

Overturf, M. L. & Loose-Mitchell, D. S. (1992) *In vivo* model systems: the choice of experimental animal model for the analysis of lipoproteins and atherosclerosis. *Current Opinion in Lipidology*, **3**, 179–85.

Reddy, J. K. & Lalwani, N. D. (1987) Carcinogenesis by hepatic peroxisome proliferators: evaluation of the risk of hypolipidemic drugs and industrial plasticisers to humans. *CRC Critical Reviews in Toxicology*, **12**, 1–58.

Richmond, W. (1992) Analytical reviews in clinical biochemistry: the quantitative analysis of cholesterol. *Annals of Clinical Biochemistry*, **29**, 577–97.

Rifai, N., Warnick, G. R., McNamara, J. R., Belcher, J. D., Grinstead, G. F. & Frantz, I. D. (1992) Measurement of low-density-lipoprotein cholesterol in serum: a status report, *Clinical Chemistry*, **38**, 150–60.

Ross, R. (1986) The pathogenesis of atherosclerosis. *New England Journal of Medicine*, **314**, 488–500.

Sampson, M., Ruddel, M. & Elin, R. J. (1994) Effects of specimen turbidity and glycerol concentration on nine enzymatic methods for triglyceride determination. *Clinical Chemistry*, **40**, 221–6.

Shepherd, J. & Packard, C. J. (1988) Pharmacological approaches to the modulation of plasma cholesterol. *Trends in Pharmacological Science*, **9**, 326–9.

Sjoblom, L. & Eklund, A. (1989) Determination of HDL_2 cholesterol by precipitation with dextran sulphate and magnesium chloride: establishing optimal conditions for rat plasma. *Lipids*, **24**, 532–4.

Temmerman, A. M., Vonk, R. J., Niezen-Koning, K., Berger, R. & Fernandes, J. (1989) Effects of dietary cholesterol in the mongolian gerbil and the rat: a comparative study. *Laboratory Animals*, **23**, 30–5.

Tschantz, J-C. & Sunahara, G. I. (1993) Microaffinity chromatographic separation and characterisation of lipoprotein fractions in rat and mongolian gerbil serum. *Clinical Chemistry*, **39**, 1861–7.

Tuten, T., Robinson, K. A. & Sgoutas, D. S. (1993) Discordant results for triglycerides in pig sera. *Clinical Chemistry*, **39**, 125–8.

Van Zutphen, L. F. M. & den Bieman, M. G. C. W. (1981) Cholesterol response in inbred strains of rats, *Rattus norvegicus*. *Journal of Nutrition*, **111**, 1833–8.

Warnick, G. R. & Albers, J. J. (1978) A comprehensive evaluation of the heparin-manganese precipitation procedure for estimating high density lipoprotein cholesterol. *Journal of Lipid Research*, **19**, 65–76.

Watanabe, Y. (1980) Serial inbreeding of rabbits with hereditary hyperlipidemia (WHHL-rabbit). *Atherosclerosis*, **36**, 261–8.

Watson, T. D. G. & Barrie, J. (1993) Lipoprotein metabolism and hyperlipidaemia in the dog and cat: a review. *Journal of Small Animal Practice*, **34**, 479–87.

Weingand, K. W. & Hudson, C. L. (1989) Accurate measurement of total plasma triglyceride concentrations in laboratory animals. *Laboratory Animal Science*, **39**, 453–4.

Weingand, K. W. & Daggy, B. P. (1991) Effects of dietary cholesterol and fasting on hamster plasma lipoprotein lipids. *European Journal of Clinical Chemistry and Clinical Biochemistry*, **29**, 425–8.

Wiebe, D. A. & Bernert, J. T. (1984) Influence of incomplete cholesteryl ester hydrolyis on enzymic measurements of cholesterol. *Clinical Chemistry*, **30**, 352–6.

Zimetbaum, P., Frishman, W. H. & Kahn, S. (1991) Effects of Gemfibrozil and other fibric acid derivatives on blood lipids and lipoproteins. *Journal of Clinical Pharmacology*, **31**, 25–37.

14

Proteins

G. O. EVANS

Although there are more than 3000 distinct proteins, in most laboratories performing toxicological studies, protein measurements are confined to plasma (or serum) total protein, albumin, calculated globulin and albumin:globulin ratios. In addition, total protein in urine is measured by qualitative test strips (dipsticks) or quantitative methods. These measurements can be supported by qualitative or quantitative assessment of broad protein fractions following electrophoretic separation. Far fewer laboratories measure specific proteins, although these measurements are commonly used in the diagnosis of human disease. Occasionally, protein measurements may be useful in other body fluids such as cerebrospinal fluid, peritoneal fluids and saliva.

Many plasma proteins including albumin and most of the globulins are synthesized in the liver, while the immunoglobulins are synthesized in the plasma cells and B-lymphocytes of the spleen, bone marrow and lymph nodes.

The protein molecular masses vary from 65 kDa for albumin to 2750 kDa for beta-lipoprotein, and the molecular size of the individual protein is an important factor in determining its distribution and transport by active and passive mechanisms. Some proteins move freely in the extracellular and intravascular spaces, while other intracellular proteins are released only after cell damage.

Some of the major plasma proteins are listed in Table 14.1 together with the broad protein fractions designated by electrophoretic mobilities. Proteins can also be divided into several broad functional categories, but several proteins have more than one of these functions. These categories are acute phase (reactant) proteins, transport proteins, immunoproteins, complement proteins and coagulation proteins. The enzymic roles of proteins are discussed in other chapters (see Chapter 5 – General Enzymology).

Few protein changes are pathognomonic and small changes of individual proteins of the globulin fractions are often masked by opposing changes in the other protein fractions. In plasma, the total protein varies from approximately 50 to 70 $g.L^{-1}$ depending on the species, with slightly more than half of the total being albumin: thus, any major change of plasma albumin will be reflected by the total protein values. The metabolic rates for individual proteins vary in the different

Table 14.1 Some plasma proteins of interest

Prealbumin (transthyretin)
Albumin
alpha 1 globulins
alpha 1 Antitrypsin
alpha 1 Acid glycoprotein (orosomucoid)
alpha Lipoproteins
alpha 2 globulins
alpha 2 Macroglobulin
Caeruloplasmin (ceruloplasmin, copper oxidase)
Haptoglobin
beta globulins
C-reactive protein
Amyloid A
Complement, C3 and C4
Transferrin (siderophilin)
Ferritin
beta Lipoprotein
Fibrinogen
gamma globulins
Immunoglobulins

species, with generally faster rates in smaller laboratory animals (Table 14.2), so proteinic changes due to xenobiotics (as with enzymes) can occur more rapidly in smaller animals. Table 14.3 lists some of the common causes for change of plasma protein values. A rise in plasma albumin is usually due to dehydration, and this can be confirmed by associated increases of plasma globulins, blood haemoglobin and packed cell volume (haematocrit). Severe hypoproteinaemia can be associated with oedema and ascites due to the major osmotic influence of albumin.

14.1 Acute Phase (Reactant) Proteins

Inflammation is accompanied by a variety of physiological and biochemical changes which are collectively called the acute phase response. As part of this response, hepatic synthesis of approximately 30 plasma proteins changes: these are the acute phase proteins. Some of these proteins are produced in increased amounts, i.e. positive, while the production of other proteins is diminished, i.e. the negative acute phase proteins. Given these bidirectional effects, it is not surprising that measurements of plasma total protein and globulin often fail to detect changes due to inflammation (Weimer *et al.*, 1972; Nakagawa *et al.*, 1984). The acute plasma protein patterns also varied in three models of hepatotoxicity (Fouad *et al.*, 1983). Reliance on a single acute phase protein measurement as an indicator of inflammation is not advisable, as the response times of these proteins are variable in both acute and chronic inflammatory conditions (Ganrot, 1973; Lewis *et al.*, 1989). Stress can also affect levels of acute phase proteins (van Gool *et al.*, 1990).

Table 14.2 Estimated half-lives (T½ in days) for albumin and immunoglobulins in some species

Species	Albumin	Immunoglobulins
Mouse	1.9	5
Rat	2.5	5
Guinea pig	2.8	5 to 7
Rabbit	5.7	7 to 9
Dog	8.2	8

Table 14.3 Causes of hyper- and hypoproteinaemia

Hypoproteinaemia
- Overhydration
- Increased protein loss
  - in urine
  - in gastrointestinal fluid
  - haemorrhage
  - parasitic infection
  - phlogistic response
- Decreased protein synthesis
  - protein starvation
  - malabsorption
  - hepatotoxicity
  - congestive heart failure

Hyperproteinaemia
- Dehydration
- Inflammation
- Hepatotoxicity
- Primary glomerular disease
- Neoplasia
- Viral, protozoal and fungal infections

There are marked differences between species for acute phase proteins, particularly for C-reactive protein, amyloid protein and alpha-2-macroglobulin (Kushner and Mackiewicz, 1987; Eckersall *et al.*, 1991). Plasma amyloid protein is the major acute protein in the mouse (Pepys *et al.*, 1979) while alpha-2-macroglobulin is the major acute phase protein in the rat, and there is a protein similar to CRP in female hamsters (Coe and Ross, 1983).

14.2 Immunoproteins

The immune system can be subdivided into a non-specific system involving factors such as lysozyme, complement and interferons, natural killer cells and phagocytes, and a second system involving the immunoglobulins and the T-lymphocytes. This second system is triggered by incomplete actions of non-specific factors, with the immunoglobulins having the functions of antigen recognition and subsequent initiation of effector mechanisms to destroy or nullify these antigens. There are five primary immunoglobulins (Ig) isotypes (or classes): IgA, IgM, IgG, IgD and IgE, and in laboratory animals they share some common structural features (Neoh *et al.*, 1973). The immunoglobulins can be further divided into subclasses in several species including the mouse, rat and dog. In addition to any of the direct actions on the immune system, the levels of immunoglobulins which are found in the plasma, lymph and other body fluids may be altered by any general effect on the rate of protein synthesis or clearance, such as protein-losing enteropathy, nephrotoxicity or hepatotoxicity.

The expansion of immunology and the increasing recognition of immune responses to xenobiotics have led to immunotoxicology developing as a distinct discipline (Burrell *et al.*, 1992). Tiered approaches for evaluating the immune system have been proposed and this includes the measurement of immunoglobulins, despite the lack of recognized, standardized and validated procedures (Vos, 1980; Luster *et al.*, 1988). Xenobiotics may act as immunosuppressants, immunostimulants or immunogens, i.e. the xenobiotic is coupled to a carrier protein or other molecule and induces antibody formation. Several investigators have reported on the application of immunoglobulin measurements in toxicology (Descotes and Mazue, 1987; van Loveren and Vos, 1989), and it appears to be an area for future development.

The complement proteins constitute about 12% of the total globulin protein fraction, with differences occurring between species both in structure and concentration (Gigli and Austen, 1971; Barta, 1984). There are at least 20 immunologically distinct components which are synthesized in the liver and these proteins therefore can be affected by hepatotoxicity. Although total complement activity can be measured by haemolytic assays, large reductions of complement proteins are required for this test to be abnormal. Complement proteins are not commonly measured in toxicology studies.

14.3 Coagulation Proteins

Haematology laboratories are generally given the responsibilities for measuring these proteins, although the use of chromogenic substrates with chemistry analysers has led to an increasing involvement for clinical chemists. For regulatory toxicology studies, coagulation measurements are usually limited to prothrombin time and activated partial thromboplastin time measurements, which are indicative for deficiencies in the extrinsic and intrinsic coagulation pathways (Theus and Zbinden, 1984). The coagulation proteins synthesized by the liver can be useful markers of hepatotoxicity (Pritchard *et al.*, 1987). Fibrinogen which has a short half-life of between 1 and 4 days in the different species acts as an acute phase protein in addition to its role in coagulation. Coagulation protein measurements are of particular interest when testing compounds deliberately designed to affect fibrinolysis.

14.4 Transport Proteins

The major transport proteins include prealbumin, albumin, acid glycoprotein, haptoglobin, transferrin, caeruloplasmin, lipoproteins, thyroid hormone binding proteins and corticosteroid binding globulin: the last three groups of transport proteins are discussed in Chapters 8 and 13. The binding of xenobiotics to albumin and acid glycoprotein is important for drug distribution, and marked changes of drug binding protein can alter the apparent blood concentration of test compound and thus resultant toxicity (Goldstein *et al.*, 1974). Albumin binds to calcium and magnesium, and reduced plasma albumin levels lead to reduced levels of these cations in plasma. Haptoglobin measurements can be used to determine the presence of intravascular haemolysis in a number of species (Walker *et al.*, 1991), and transferrin can be used to monitor changes of iron metabolism.

Other applications for plasma protein measurements include acute phase proteins in contact sensitivity studies (Kimber *et al.*, 1989), immune complexes in renal pathology, and where transgenic animal models of immunodeficiency are used. As a general rule, measurement of proteins associated with malignancy, for example in long-term rodent studies, does not help risk assessment (Dolezalova *et al.*, 1983). Glycosylated haemoglobin and albumin assays may be performed in toxicological studies of hypoglycaemic agents or where there is evidence of acquired hyperglycaemia (Higgins *et al.*, 1982; Neuman *et al.*, 1994).

14.5 Urinary Proteins

These measurements and their diagnostic value are discussed in Chapter 7 on Nephrotoxicity. There are some important species differences particularly for rodents: a family of prealbumins known as the major urinary protein (MUP) is found in the mouse (Finlayson and Baumann, 1958), and an alpha 2_u-globulin is the major protein in male rat urine. Generally male rodents are more proteinuric, and there are marked changes with age, with increased urinary albumin in the aged rat. An important side issue is the level of IgG anti-rat antibodies in some affected individuals when considering human allergy to laboratory animals (Botham *et al.*, 1989).

14.6 Preanalytical Factors

In addition to the information provided in Chapter 3 on preanalytical factors, several effects on plasma proteins have been described. Examples include strain and sex differences for mouse prealbumins (Reuter *et al.*, 1968), and age-related changes in the hamster (House *et al.*, 1961), dog (Uchiyama *et al.*, 1985) and the rat (Wolford *et al.*, 1987). Circadian rhythms from plasma proteins have been described in rats (Scheving *et al.*, 1968).

For rats and other small rodents, the site and anaesthetic agent used for blood collection have marked effects on plasma protein values (Neptun *et al.*, 1985). Ketamine anaesthesia alters plasma protein levels in rhesus monkeys (Bennett *et al.*, 1992). Dietary effects on plasma proteins have been reported in short-term studies (Maejima and Nagase, 1991) and long-term studies (Pickering and Pickering, 1984). Colvin and Lee-Wang (1974) reported dietary protein altered the effects of warfarin on coagulation and other plasma proteins.

14.7 Measurements of Proteins

Plasma protein status is generally assessed by measuring total protein and albumin, before proceeding to qualitative or quantitative separation of proteins by various electrophoretic techniques or specific protein determinations. Plasma protein values are slightly higher (approximately 5%) than corresponding serum values due to the presence of fibrinogen, although cold storage can cause this fraction to precipitate. Globulin values are usually determined by subtracting albumin values from total protein, and the plasma globulin value will also include a contribution due to fibrinogen. Methods exist for direct measurement of globulin, e.g. using glyoxylic acid, but are rarely used. Changes of plasma albumin or globulin fractions are also reflected by alterations of the calculated albumin:globulin (A:G) ratio.

The older flocculation tests used with serum to indicate changes in the proportions of different proteins, e.g. thymol turbidity test, should be discarded and replaced by more appropriate quantitative methods. Another non-specific test is the erythrocyte sedimentation rate (ESR) which alters in response to changes of acute phase proteins, immunoglobulins, fibrinogen and albumin.

The calibration of total protein, albumin and electrophoretic and specific protein methods remains a problem when analysing samples from laboratory animals. Many calibration materials are bovine or human in origin, and there remains a distinct lack of suitable reference proteins (Whicher, 1994). For most species where protein fractions are available these are often less than 95% pure. The investigator therefore needs to consider how best to show relative protein changes when designing a study: sometimes it may be necessary in the absence of absolute standards to express results as a percentage of the control group or pretreatment values.

For total protein measurements the majority of laboratories use the biuret method, which is based on the formation of a cupric ion–peptide complex in an alkaline solution. Other methods using Coomassie Blue G-250, bicinchoninic acid, or the older Lowry reagents are generally reserved for measuring the lower protein levels found in body fluids other than plasma. Several methods have been compared for human urines, and the Coomassie Blue methods appear to be the most promising for urinary protein (Evans and Parsons, 1986).

Measurements of total protein by refractometry depend on the refractive indices of plasma, and clinical refractometers are generally calibrated using human samples which have different levels of dissolved solids: results are not therefore as accurate as those obtained with the biuret method but may still be adequate in a clinical situation. Elevated plasma lipid levels interfere with protein measurements made by refractometers.

The bromocresol (or bromcresol:BCG) green dye-binding method for plasma albumin appears to be suitable for most laboratory animal species unlike some other albumin-binding dyes (Witiak and Whitehouse, 1969; Metz and Schutze, 1975; Evans and Parsons, 1988). Bromocresol purple appears to be unsuitable for most laboratory species. The bromocresol green methods are said to overestimate values below 20 g/L, but this depends on the reagent constitution, reaction conditions and more importantly on the timing of absorbance measurements. Proteins other than albumin such as alpha-1-globulins may react with bromocresol green if the absorbance measurements are made more than 60 s after the reagent

Table 14.4 Qualitative and quantitative methods for measuring proteins

Electrophoretic methods for groups of proteins
(Various support media, buffers, voltage and amperage and times; SPE = serum protein electrophoresis)
Agarose gel
Cellulose acetate
Starch gel
Polyacrylamide gel
Two-dimensional immunoelectrophoresis
Crossed immunoelectrophoresis
Immunoelectrophoretic fixation (IEF)
Methods for specific proteins
Single radial immunodiffusion (SRID)
Laurell immunoelectrophoresis ('rockets')
Immunoturbidometry
Immunonephelometry
Column chromatography, e.g. for haptoglobin
Radioimmunoassay
Enzyme-linked immunosorbent assay (ELISA)

addition (Gustafsson, 1976). Two investigators have reported problems when using bromocresol green to measure rabbit serum albumin, but others have not confirmed this finding (Fox, 1989; Hall, 1992; Evans, 1994).

Plasma protein electrophoresis using various matrices can provide more detailed information on changes of plasma protein, and there is a wide variety of available techniques (Table 14.4). Simple electrophoretic techniques using cellulose acetate or agarose support media can be used to assess changes of albumin and globulin fractions, although the support media and electrophoretic conditions may affect the electrophoretograms (Allchin and Evans, 1986) and it is sometimes necessary to adjust the electrophoretic methods for particular species if satisfactory separations are to be obtained. Using serum rather than plasma avoids complications due to fibrinogen or fibrin. The commonest dyes used with these simple techniques are Amido Black or Ponceau S, but neither of these dyes are uniformly bound by the various protein fractions so that in most cases the results are semi-quantitative. Sometimes blurring or close migration of particular plasma protein fractions prevents satisfactory quantification.

Although albumin is usually the dominant electrophoretic band, there are marked species differences for the electrophoretic patterns. An additional complication is the description of protein electrophoretic mobilities compared with the migration of human proteins in the same electrophoretic system, rather than by using appropriate protein markers for each species. Several individual proteins occur within the same electrophoretically separated fraction, and it is therefore difficult to detect an increase of a specific protein. Statistically significant differences between treatment groups should be interpreted with due consideration of the imprecision of the electrophoretic methods, particularly for fractions which are less than 10% of the total protein concentration.

The inadequacies of these simpler one-dimensional techniques have led investigators to use techniques such as two-dimensional immunoelectrophoresis, where it is possible to demonstrate additional changes of plasma proteins, e.g. in rats given hypolipidaemic agents (Hinton *et al.*, 1985). When quantitative immunoelectrophoretic techniques were used to study the acute phase response to carrageenan in rats, changes for over 20 individual proteins were demonstrated (Weimer *et al.*, 1972), in contrast to the simple observations of increased plasma alpha- and beta-globulins made with cellulose acetate electrophoretic methods (Scherer *et al.*, 1977).

The application of qualitative and quantitative immunochemical protein measurements is limited by the lack of available antisera suitable for all species, although this situation is improving. Antisera against human proteins are often useful for closely related species, and some phylogenetically distant species have similar epitopes on their corresponding proteins which allows protein measurements in these species (Brun *et al.*, 1989; Hau *et al.*, 1990; Salauze *et al.*, 1994). Some proteins require species-specific antisera, e.g. CRP (Balz *et al.*, 1982; Eckersall *et al.*, 1991). Where a suitable antiserum is available, it is still necessary to test that the concentrations of the antiserum and antigen are appropriate: incorrect reagents may lead to over- or underestimation for the protein under examination.

14.7.1 Urine Protein

Several test strips or dipsticks are commercially available and commonly used to assess proteinuria, but there are several limitations to their use. Highly alkaline or buffered urines give false positive protein reactions and the test strips are primarily sensitive to changes in albumin and not globulin fractions (Evans and Parsons, 1986). Allowances should be made for the urine volume and concentration when interpreting test strip results. Various methods used for assessing plasma proteins can be adapted for urinary protein examinations (see earlier section on total protein measurements and Chapter 7 – Nephrotoxicity).

14.7.2 Proteins in other Body Fluids

The fluids which sometimes accumulate in the peritoneum and pleural cavities vary in their protein content. They may be ultrafiltrates with low total protein and particularly high molecular mass protein, or the fluids may contain significant amounts of protein including immunoglobulins, which may reflect malignancy or infection. Simple protein measurements including electrophoresis may provide useful information on the nature and origin of these fluids.

References

Allchin, J. P. & Evans, G. O. (1986) Serum protein electrophoretic patterns of the marmoset, *Callithrix jacchus*. *Journal of Comparative Pathology*, **96**, 349–52.

Barta, O. (1984) *Laboratory Techniques of Veterinary Clinical Immunology*. Springfield, IL: Charles C. Thomas.

Balz, M. L., deBeer, F. C., Feinstein, A., Munn, E. A., Milstein, C. P., Fletcher, T. C., March, J. F., Taylor, J., Brunton, C., Clamp, J. R., Davies, A. J. S.

& PEPYS, M. B. (1982) Phylogenetic aspects of C-reactive protein and related proteins. *Annals of the New York Academy of Science*, **389**, 49–73.

BENNETT, J. S., GOSETT, K. A., MCCARTHY, M. P. & SIMPSON, E. D. (1992) Effects of ketamine hydrochloride on serum biochemical and hematologic variables in Rhesus monkeys (*Macaca mullata*). *Veterinary Clinical Pathology*, **21**, 15–18.

BOTHAM J. W., MCCALL, J. C., TEASDALE, E. L. & BOTHAM, P. A. (1989) The relationship between exposure to rats and antibody production in man: IgG antibody levels to rat urinary protein. *Clinical and Experimental Allergy*, **19**, 437–41.

BRUN, E., LINGAAS, F., LARSEN, H. J. (1989) Turbidimetric measurement of immunoglobulin G in serum using an automatic centrifugal analyser. *Research in Veterinary Science*, **46**, 168–71.

BURRELL, R., FLAHERTY, D. K. & SAUERS, L. J. (1992) *Toxicology of the Immune System: A Human Approach*. New York: Van Nostrand Reinhold.

COE, J. E. & ROSS, M. J. (1983) Hamster female protein. A divergent acute phase protein in male and female Syrian hamsters. *Journal of Experimental Medicine*, **157**, 1421–33.

COLVIN, H. W. & LEE-WANG, W. (1974) Toxic effects of warfarin in rats fed different diets. *Toxicology and Applied Pharmacology*, **28**, 337–48.

DESCOTES, G. & MAZUE, G. (1987) Immunotoxicology. *Advances in Veterinary Science and Comparative Medicine*, **31**, 95–119.

DOLEZALOVA, V., STRATIL, P., SIMICKOVA, M., KOCENT, A. & NEMECEK, R. (1983) α-Fetoprotein (AFP) and α-2-macroglobulin (α_2-M) as the markers of distinct responses of hepatocytes to carcinogens in the rat: carcinogenesis. *Annals of the New York Academy of Science*, **417**, 294–307.

ECKERSALL, P. D., CONNER, J. G. & HARVIE, J. (1991) An immunoturbidometric method for C-reactive protein. *Veterinary Research Communications*, **15**, 17–24.

EVANS, G. O. (1994) Plasma albumin measurement in New Zealand White rabbits. *World Rabbit Science*, **1**, 25–7.

EVANS, G. O. & PARSONS, C. E. (1986) Potential errors in the measurement of total protein in male rat urine using test strips. *Laboratory Animals*, **20**, 27–31.

(1988) A comparison of two dye binding methods for the determination of dog, rat and human plasma albumin. *Journal of Comparative Pathology*, **98**, 453–60.

FINLAYSON, J. S. & BAUMANN, C. A. (1958) Mouse proteinuria. *American Journal of Physiology*, **192**, 69–72.

FOUAD, F. M., GOLDBERG, M. & RUHENSTROTH-BAUER, G. (1983) Plasma protein determination as a clinical probe for liver injury in rats induced by thioacetamide, alloxan or Ixoten. *Journal of Clinical Chemistry and Clinical Biochemistry*, **21**, 203–8.

FOX, R. R. (1989) The rabbit. In Loeb, W. F. & Quimby, F. W. (eds), *The Clinical Chemistry of Laboratory Animals*, pp. 45–6. New York: Pergamon Press.

GANROT, K. (1973) Plasma protein response in experimental inflammation in the dog. *Research and Experimental Medicine*, **161**, 251–61.

GIGLI, I. & AUSTEN, K. F. (1971) Phylogeny and function of the complement system. *Annual Review of Microbiology*, **25**, 309–32.

GOLDSTEIN, A., ARONOW, L. & KALMAN, S. M. (1974) *Principles of Drug Action*, pp. 158–64. New York: J. Wiley and Sons.

GUSTAFSSON, J. E. C. (1976) Improved specificity of serum albumin determination and estimation of 'acute phase reactants' by use of bromcresol green reaction. *Clinical Chemistry*, **22**, 616–22.

HALL, R. E. (1992) Clinical pathology of laboratory animals. In Gad, S. C. & Chengelis, C. P. (eds), *Animal Models in Toxicology*. New York: Marcel Dekker.

HAU, J., NILSSON, M., SKOVGAARD-JENSEN, H. J., DE SOUZA, A., ERIKSEN, E. & WANDALL, L. T. (1990) Analysis of animal serum proteins using antisera against human analagous proteins. *Scandinavian Journal of Laboratory Animal Science*, **17**, 3–7.

Higgins, P. J., Garlick, R. L. & Bunn, H. F. (1982) Glycosylated hemoglobin in human and animal red cells. Role of glucose permeability. *Diabetics*, **31**, 743–8.

Hinton, R. H., Price, S. C., Mitchell, F. E., Mann, A., Hall, D. E. & Bridges, J. W. (1985) Plasma protein changes in rats tested with hypolipidaemic drugs and phthalate esters. *Human Toxicology*, **4**, 261–71.

House, E. L., Pansky, B. & Jacobs, M. S. (1961) Age changes in blood of golden hamster. *American Journal of Physiology*, **200**, 1018–22.

Kimber, I., Ward, R. K., Shepherd, C. J., Smith, M. N., McAdam, K. P. W. & Raynes, J. G. (1989) Acute-phase proteins and the serological evaluation of experimental contact sensitivity in the mouse. *International Archives of Allergy and Applied Immunology*, **89**, 149–55.

Kushner, I. & Mackiewicz, A. (1987) Acute phase proteins as disease markers. *Disease Markers*, **5**, 1–11.

Lewis, E. J., Bishop, J. & Cashin, C. H. (1989) Automated quantification of rat plasma acute phase reactants in experimental inflammation. *Journal of Pharmacological Methods*, **21**, 183–94.

Luster, M. I., Munson, A. E., Thomas, P. T., Holsapple, M. P., Fenters, J. D., White, K. I., Lauer, L. D., Germolec, D. R., Rosenthal, G. J. & Dean, J. H. (1988) Development of a testing battery to assess chemical-induced immunotoxicity: national toxicology program's guidelines for immunotoxicity evaluation in mice. *Fundamental and Applied Toxicology*, **10**, 2–19.

Maejima, K. & Nagase, S. (1991) Effect of starvation and refeeding on the circadian rhythms of hematological and clinico-biochemical values and water intake of rats. *Experimental Animals*, **40**, 389–93.

Metz, A. & Schutze, A. (1975) Vergleichende untersuchungen zur Bestimmung von Gesamtprotein und Albumin im Serum von Mensch, Affe, Hund und Ratte. *Zeitschrift für Klinische Chemie und Klinische Biochemie*, **13**, 423–6.

Nakagawa, H., Watanabe, K. & Tsurufuji, S. (1984) Changes in serum and exudate levels of functional macroglobulins and anti-inflammatory effect of alpha 2-acute-phase-macroglobulin on carrageenin-induced inflammation in rats. *Biochemical Pharmacology*, **33**, 1181–6.

Neoh, S. N., Jahoda, D. M., Rowe, D. S. & Voller, A. (1973) Immunoglobulin classes in mammalian species identified by cross reactivity with antisera to human immunoglobulin. *Immunochemistry*, **10**, 805–13.

Neptun, D. A., Smith, C. N. & Irons, (1985) Effect of sampling site and collection method on variations in baseline clinical pathology parameters in Fischer 344 rats. I. Clinical Chemistry. *Fundamental and Applied Toxicology*, **5**, 1180–5.

Neuman, R. G., Hud, E. & Cohen, M. P. (1994) Glycated albumin: a marker of glycaemic status in rats with experimental diabetes. *Laboratory Animals*, **28**, 63–9.

Pepys, M. B., Baltz, M., Gomer, K., Davies, A. J. S. & Doenhoff, M. (1979) Serum amyloid P-component is an acute-phase reactant in the mouse. *Nature*, **278**, 259–61.

Pickering, R. G. & Pickering, E. C. (1984) The effects of reduced dietary intake upon the body and organ weights, and some clinical chemistry and haematological variates of the young Wistar rat. *Toxicology Letters*, **21**, 271–7.

Pritchard, D. H., Wright, M. G., Sulsh, S. & Butler, W. H. (1987) The assessment of chemically-induced liver injury in rats. *Journal of Applied Toxicology*, **7**, 229–36.

Reuter, A. M., Kennes, F., Leonard, A. & Sassen, A. (1968) Variations of the prealbumin in serum and urine of mice, according to strain and sex. *Comparative Biochemistry and Physiology*, **25**, 921–8.

Salauze, D., Serre, V. & Perrin, C. (1994) Quantification of total IgM and IgG levels in rat sera by a sandwich ELISA technique. *Comparative Haematology International*, **4**, 30–3.

SCHERER, R., ABD-EL-FATTAH, M. & RUHENSTROTH-BAUER, G. (1977) Some applications of quantitative two-dimensional immunoelectrophoresis in the study of the systemic acute-phase reaction of the rat. In Willoughby, D. A., Giroud, J. P. & Velo, G. P. (eds), *Perspectives in Inflammation: Future Trends and Developments*, pp. 437–44. Lancaster: MTP Press.

SCHEVING, L. E., PAULY, J. E. & TSAI, T-H. (1968) Circadian fluctuation in plasma proteins of the rat. *American Journal of Physiology*, **215**, 1096–101.

THEUS, R. & ZBINDEN, G. (1984) Toxicological assessment of the hemostatic system, regulatory requirements, and industry practice. *Regulatory Toxicology and Pharmacology*, **4**, 74–95.

UCHIYAMA, T., TOKOI, K. & DEKI, T. (1985) Successive changes in the blood composition of the experimental normal beagle dogs accompanied with age. *Experimental Animals*, **34**, 367–77.

VAN GOOL, J., VAN VUGT, H., HELLE, M. & AARDEN, L. A. (1990) The relation among stress, adrenalin, interleukin 6 and acute phase proteins in the rat. *Clinical Immunology and Immunopathology*, **57**, 200–10.

VAN LOVEREN, H. & VOS, J. G. (1989) Immunotoxicological considerations: a practical approach to immunotoxicity in the rat. In Dayan, A. D. & Payne, A. J. (eds), *Advances in Applied Toxicology*, pp. 143–63. London: Taylor & Francis.

VOS, J. G. (1980) Immunotoxicity assessment: screening and function. *Archives of Toxicology*, Suppl. **4**, 95–108.

WALKER, A. K., GANNEY, B. A. & BROWN, G. (1991) Haptoglobin measurements in a number of laboratory animal species. *Comparative Haematology International*, **1**, 224–8.

WEIMER, H. E., ROBERTS, D. M., VILLANUEVO, P. & PORTER, H. G. (1972) Genetic differences in electrophoretic patterns during the phlogistic response in the albino rat. *Comparative Biochemistry and Physiology*, **43B**, 965–73.

WHICHER, J. T. (1984) Functions of acute phase proteins in the inflammatory response. In Arnaud, P., Bienvenu, J. & Laurent, P. (eds), *Marker Proteins in Inflammation*, Vol. 2, pp. 90–8. New York: Walter de Gruyter.

WITIAK, D. T. & WHITEHOUSE, M. W. (1969) Species differences in the albumin binding of 2,4,6-trinitrobenzaldehyde, chlorophenoxy-acetic acids, 2-(4′-hydroxybenzeneazo)benzoic acid and some other acidic drugs – the unique behaviour of plasma. *Biochemical Pharmacology*, **18**, 971–7.

WOLFORD, S. T., SCHROER, R. A., GALLO, P. P., GOHS, F. X., BRODECK, M., FALK, H. B. & RUHREN, R. (1987) Age-related changes in serum chemistry and hematology values in normal Sprague–Dawley rats. *Fundamental and Applied Toxicology*, **8**, 80–8.

Abbreviations

AACC	American Association for Clinical Chemistry
ABPI	Association of British Pharmaceutical Industry
ACCA	Animal Clinical Chemistry Association, UK
ADI	Acceptable Daily Intake
BTS	British Toxicology Society
CAS	Chemical Abstract Service
CFR	Code of Federal Regulations, USA
CPMP	Committee for Proprietory Medicinal Products – the EEC regulatory body
CTC	Clinical Trials Certificate
CTM	Clinical Trials Material
CTX	A scheme giving exemption to full requirements of a CTC in the United Kingdom
CV	Coefficient of Variation
DoH	Department of Health, UK
DGKC	Deutsche Gessellschaft für Klinische Chemie
EDI	Estimated Daily Intake
EEC	European Economic Community
ELISA	Enzyme-linked immunosorbent assay
EPA	Environmental Protection Agency, USA
EQAS	External Quality Assessment Scheme
FASEB	Federation of American Societies for Experimental Biology
FDA	Food and Drugs Administration, USA
FHSA	Federal Hazardous Substances Act, USA
FIFRA	Federal Insecticide Fungicide and Rodenticide Act, USA
GCP	Good Clinical Practice
GLP	Good Laboratory Practice
GMP	Good Manufacturing Practice
IARC	International Agency for Research on Cancer
IFCC	International Federation of Clinical Chemistry
IND	Investigational New Drug

IUB	International Union of Biochemistry
IUPAC	International Union of Pure and Applied Chemistry
JPMA	Japanese Pharmaceutical Manufacturers Association
LOEL	Lowest observed effect level
MAA	Marketing Authorisation Application. A request to government for permission to market a product
MCA	Medicine Control Agency – Regulatory body, UK
MTD	Maximum Tolerated Dose
NCE	New Chemical Entity
NDA	New Drug Application
NOAEL	No observed adverse effect level
NOEL	No observed effect level
NVKC	Nederlandse Vereniging voor Klinische Chemie
OECD	Organisation for Economic Cooperation and Development
OTC	Over-the-counter sales without prescription
RIA	Radioimmunoassay
SAR	Structure–Activity Relationship
SD	Standard Deviation
SE	Standard Error
SFBC	Société Français de Biologie Clinique
QA	Quality Assurance
QC	Quality Control

SI Units and Conversion Tables

The International System of Units (Système International; SI) was adopted in 1960 by the General Conference of Weights and Measures as a coherent system based on seven basic units: the metre, kilogram, second, ampere, kelvin, candela and mole. The following table lists the units recommended for some of the common biochemical measurements, together with factors for converting the SI values from the traditional non-SI units. Although it was anticipated that the system would be adopted universally, this is not the case.

References

BARON, D. N., BROUGHTON, P. M. G., COHEN, M., LANSLEY, T. S., LEWIS, S. M. & SHINTON, N. K. (1974) The use of SI units in reporting results obtained in hospital laboratories. *Journal of Clinical Pathology*, **27**, 590–7.

DYBKAER, R. (1979) Lists of quantities in Clinical Chemistry. Approved recommendation (1978). IUPAC AND IFCC. *Clinica Chimica Acta*, **96**, 185–204F.

LAPOSATA, M. (1992) *SI Conversion Guide*. Waltham MA, USA: Massachusetts Medical Society.

EDITORIAL (1986) Now read this: the SI units are here. *Journal of the American Medical Association*, **255**, 2329–39.

Other Units and Conversion Factors

1 To convert from traditional (or conventional) units to SI units, the following formulae can be used:

Mass unit/dL × 10 = mass units/L

e.g. total protein and protein fractions

or for mass concentration to substance concentration

$$\frac{\text{Mass unit/dL}}{\text{Molecular mass}} \times 10 = \text{substance concentration mmol/L}$$

e.g. glucose, cholesterol

Table I SI units and conversion factors

Analyte	SI unit	Traditional unit	Conversion factor to SI
Albumin	g/L	g/dL	10
	μmol/L	g/dL	149
Bicarbonate	mmol/L	meq/L	1
Total bilirubin	μmol/L	mg/dL	17.1
Calcium	mmol/L	mg/dL	0.25
		meq/L	0.5
Chloride	mmol/L	mg/dL	1
Cholesterol	mmol/L	mg/dL	0.0259
Creatinine	μmol/L	mg/dL	88.4
Globulin	g/L	g/dL	10
Glucose	mmol/L	mg/dL	0.0555
Inorganic phosphate	mmol/L	mg/dL	0.323
Iron	μmol/L	μg/dL	0.179
Magnesium	mmol/L	mg/dL	0.411
		meq/L	0.5
Potassium	mmol/L	meq/L	1
Sodium	mmol/L	meq/L	1
Thyroxine	nmol/L	μg/dL	12.9
Total protein	g/L	mg/dL	10
Tri-iodothyronine	nmol/L	ng/dL	0.015
Triglycerides	mmol/L	mg/dL	0.0113
Urate (uric acid)	μmol/L	mg/dL	59.48
Urea	mmol/L	mg/dL	0.166

These conversion factors apply to serum or plasma. For some substances a different factor will need to be used, e.g. for urinary creatinine a factor of 8.84 will convert g/L to mmol/L.

2 **pH/Hydrogen ion**
The recommended SI unit is nmol/L. This relationship is expressed by the formula:

$$pH = -\log_{10}[H^+]$$

where $[H^+]$ is expressed in mol/L.

3 **Blood urea nitrogen**
To convert mg/dL to mmol/L, multiply by 0.357.
To convert urea nitrogen values mg/dL, multiply by 2.14 to obtain urea mg/dL values.

4 **Molarity and molality**
The mole is a unit of mass and is expressed as the molecular weight of a substance in grammes (g). A **molar** solution will contain one mole solute per L of solution. A **molal** solution contains one mole solute per 1000 g of solvent.

5 **Osmolarity and osmolality**
Osmolarity refers to the molar concentration (moles/L), whereas osmolality

refers to molal concentrations and is usually expressed as mosmol/kg body water.

6 **Enzymes**

Enzyme values vary considerably between laboratories and these values are dependent on measurement temperature, substrate concentration, pH, buffer, activators and other methodological variations. Enzyme activities may be expressed by various units; these units include:

International Unit – defined as the enzyme aactivity which will convert one micromole of substrate per minute under defined conditions. It may be expressed as IU/L, or sometimes as U/L or mIU/ml. One IU/L corresponds to 16.67 nkatal/L.

Katal – defined as the catalytic activity that will convert one mole of substrate per second under defined conditions. One nanokatal (nKat) equals 10^{9} mol/s or 0.06 IU/L.

If the reaction conditions vary, e.g. using different substrates for the same enzyme, then using these conversion factors may be inappropriate for comparison. Even for common enzymes such as the aminotransferases there are several different national and international recommended methods for use with human sera or plasma, and these yield differing values despite conformity in units. There are no internationally recommended conditions for samples obtained from laboratory animals.

General References for Animal Clinical Chemistry Data

There is no substitute for each laboratory establishing its own reference values given the effects of preanalytical and analytical factors discussed in Chapter 3 (and other chapters). However, investigators using a method for the first time may wish to compare data with previously published data for the same species and strain. The following list includes books and papers where reference values have been published and these can be used as a guide. In addition, many of the papers cited in each chapter contain data for 'control' group animals.

Books

EVERETT, R. M. & HARRISON, S. D. (1983) Clinical biochemistry. In Foster, H. L., Small, J. D. & Fox, J. G. (eds), *The Mouse in Biomedical Research*, pp. 313–26. New York: Academic Press.

KANEKO, J. J. (1989) *Clinical Biochemistry of Domestic Animals*, 4th Edn. San Diego: Academic Press.

LAIRD, C. W. (1974) In Melby, E. C. & Altman, N. H. (eds), *Handbook of Laboratory Animal Science*, vol. 2, pp. 347–436. Cleveland: CRC Press.

LOEB, W. F. and QUIMBY, F. W. (eds) (1989) *The Clinical Chemistry of Laboratory Animals*. New York: Pergamon Press.

MCLAUGHLIN, R. M. & FISH, R. E. (1994) Clinical biochemistry and hematology. In Manning, P. J., Ringler, D. H. & Newcomer, C. E. (eds) *The Biology of the Laboratory Rabbit*, pp. 111–27. San Diego: Academic Press.

MITRUKA, B. M. & RAWNSLEY, H. M. (1977) *Clinical Biochemical and Hematological Reference Values in Normal Experimental Animals*. New York: Masson Publishing USA.

RINGLER, D. H. & BABICH, L. (1979) Hematology and clinical chemistry. In Baker, H. J., Lindsey, J. R. & Weisbroth, S. H. (eds), *The Laboratory Rat*, pp. 105–20. Florida: Academic Press.

TOMSON, F. N. & WARDROP, K. J. (1987) Clinical chemistry and hematology. In Van Hoosier, G. L. & McPherson, W. (eds), *Laboratory Hamsters*, pp. 43–59. Florida: Academic Press.

General Papers

MATSUZAWA, T., NOMURA, M. & UNNO, T. (1993) Clinical pathology reference ranges of laboratory animals. *Journal of Veterinary Medicine and Science*, **55**, 351–62.

NEPTUN, D. A., SMITH, C. A. & IRONS, R. D. (1985) Effect of sampling site and collection method on variations in baseline clinical pathology parameters in Fischer-344 rats. 1. Clinical chemistry. *Fundamental and Applied Toxicology*, **5**, 1180–5.

WOLFORD, S. F., SCHROER, R. A., GOHS, F. X., GALLO, P. P., BRODECK, M., FALK, H. B. & RUHREN, R. (1986) Reference range data base for serum chemistry and haematology values in laboratory animals. *Journal of Toxicology and Environmental Health*, **18**, 161–88.

Index

* 9 7 8 0 7 4 8 4 0 3 5 1 6 *
T - #0055 - 071024 - C0 - 254/178/12 [14] - CB - 9780748403516 - Gloss Lamination